CW00553220

Economics of the International Coal Trade

Lars Schernikau

Economics of the International Coal Trade

The Renaissance of Steam Coal

Dr. Lars Schernikau
HMS Bergbau AG
Germany

ISBN 978-90-481-9239-7 e-ISBN 978-90-481-9240-3
DOI 10.1007/978-90-481-9240-3
Springer Dordrecht Heidelberg London New York

Library of Congress Control Number: 2010929255

Printed on acid-free paper

Springer is part of Springer Science+Business Media (www.springer.com)

Foreword 1

Coal, the catalyst of the industrial age, is now poised to shape how the world consumes energy in the twenty-first century. The ascendance of oil in the global economy to a looming peak is forcing countries, companies, and consumers to reconsider their relationship to something they cannot live without: energy. And while other sources of energy, such as nuclear, natural gas, and renewables, will all play an ever greater role in serving demand, intelligent observers would be wise not to miss what is perhaps the clearest trend of all: the second coal era is now upon us.

The developing world is rapidly electrifying in order to drive economic growth. And electrification means coal. This is particularly true in India and China, where coal offers the cheapest and most reliable route to electric power. These two countries alone will drive 80% of coal consumption growth to 2030. The International Energy Agency expects that global coal consumption will increase by 60% in the next two decades. Lars Schernikau's *The Renaissance of Steam Coal* could not come at a better time. Lars' skill in tracing the long arc of industrial evolution paired with his acute knowledge of the coal market make his arguments both insightful and highly credible. He begins by succinctly framing the problem and distilling our current predicament: we are caught between 'the Oil Age' and the 'the Solar Age' (by this he means the age of renewables). We are all optimists to some degree, but the realists among us know that renewable energy is not going to completely fill this gap for decades. Where, then, does this leave us? It leaves us with *The Renaissance of Steam Coal.*

Coal is now the world's fastest growing source of fossil fuel, a position it is expected to hold for the foreseeable future. Yet the coal market is far less well understood than the oil or gas markets. Academics, policymakers, and market participants are faced with the prospect of our collective knowledge about this market not keeping pace with its increasing relevance. Coal, only 30 years ago a localized fuel source, is now a volatile global commodity with banks and hedge funds piling into the once traditional business. Not only is the coal market now more relevant than ever, it's also much more complex.

As a shareholder of HMS Bergbau AG, a deeply experienced European coal trading firm, Lars is in a unique position to take us inside this world. As the global trade grows, growth is likely to be uneven. Major exporters such as Indonesia,

South Africa, Colombia, and Australia face a range of issues that will either enable or constrain their ability to serve the world's growing fleet of coal-fired power plants. His WorldCoal model provides us with the economic foundation of cost curves and competitive theory upon which we can layer a nuanced treatment of issues like increased local demand reducing exports or unstable governance threatening production. The result is a picture of growth patterns in the global trade that brings to bear the set of tools that are now essential for any coal market participant: knowledge of power markets, geopolitical astuteness, a detailed understanding of mining economics, and an acquaintance with global freight markets.

The market is becoming more sophisticated as trade becomes more complicated. Coal exchanges are expanding in an environment of high credit risk. Futures, swaps, and options – once the domain of oil markets – are taking hold in the coal markets. By many estimates, the financial trade is now six times the physical coal trade in Europe. The launch of several futures contracts for the Newcastle market is about to test whether the Pacific will follow Europe's lead in the commoditization of coal. For many observers, the appearance of derivatives in Asia (and the banks right behind them) will be enough to prove that coal is decidedly headed the direction of oil. But coal is not oil. The heterogeneous nature of this fuel and the prevalence of markets with complicated logistics ensure that not all of the coal trade can go on-screen; Renaissance puts the ceiling at 50% of the internationally traded market. If this is indeed the future of coal, the market will demand players that can place a physical cargo into Guangzhou one minute and short API2 swaps the next.

It would be a near requirement to be unabashedly bullish coal if not for one critical factor: carbon dioxide. Coal use is now the leading source of anthropogenic climate change. The European Emissions Trading Scheme (EU ETS) has put the brakes on coal burn by forcing power and coal traders alike to factor European Union Allowances (EUAs) into the power price. Watching 'clean spark' and 'clean dark' spreads that track coal's competitiveness against natural gas in the merit order is now required hobby for anyone trading in Europe. America is decidedly headed the same direction, and the world aspires to replace the Kyoto Protocol by the time it expires in 2012 with a binding cap on global emissions. Whether the coal business believes in global warming or not is irrelevant to the coal trade; the politics are clear, the policies are being strengthened, and the economics are unavoidable. Coal's relationship to climate policy will be one of the determining factors in deciding who uses coal and where. This book directly confronts this issue in a novel way, arguing that geographically limited application of CO_2 pricing may drive down coal use in Europe or the United States, but this would depress world coal prices and thus incentivize even greater coal use (and thus emissions) in the coal-hungry developing world. This nuanced global view of the coal trade and its relationship to climate policy need to be considered by policymakers, environmentalists, and coal producers alike.

Coal, like it or not, is about to witness a period of phenomenal growth. Unless the global community understands the coal market on a level consummate with its importance, we cannot make good markets or good policy. *The Renaissance of Steam Coal* makes an important contribution to this end and should be read by anyone who has a stake in the future of global energy markets.

Program on Energy and Sustainable Development
Stanford University
Stanford
CA, USA

Richard K. Morse

Foreword 2

Energy, the delivery of light and power, is one of the most important and potentially problematic of contemporary issues. As developing countries electrify their communities, already modernized societies seek to feed an ever-increasing technologically dependant way of life, and global populations continue to increase, the energy needs of the world are set to skyrocket. However nobly motivated the proponents of renewable energy may be, the world's energy demands are growing faster, and with greater force, than current renewable energy technology can adequately be made available or efficient enough to cope with. Environmental-political forces may have the power to sway the direction of future energy sourcing, but for coming decades, as billions of people turn on the lights for the first time, coal-fueled power stations will, and should, be the logical stopgap to the energy needs of the global community. The author of this book, Lars Schernikau, has identified this reemerging market and the general misinformation surrounding coal-fired power generation and put together a study seeking to inform, update, and educate people about today's coal industry. Up to now, the coal mining industry has been figuratively shrouded in industrial-era black dust. Few understand the modern processes or production barriers. Misinformation, propagated for a myriad of reasons – from fiction to political ideology – pollutes our understanding of the current industry. *The Renaissance of Steam Coal* is a timely, unbiased analysis of coal's ability to fuel the growing energy market, and of what a rebirth of coal-derived power will mean for industry, investors, and the quality of the world's energy supply.

Historically, the coal industry has been associated with images populated by Charles Dickens's downtrodden characters, struggling for survival in a soot-covered London; a time before child labor laws, when mine safety consisted of little more than a canary in a cage. The coal mining industry is over 200 years old and, along with many other long-in-the-tooth trades, both does and doesn't deserve its reputation. Underground mining, if not undertaken in a professional and safety-conscious manner, can result in significant fatalities. Almost every known form of premature death can occur in poorly managed underground mining operations. Burning coal in open fires or in boilers that do not have pollutant removal systems produces a variety of emissions with various environmentally damaging effects, ranging from carcinogens to acid-rain-forming sulfur compounds, as well as visual pollution in the form of the notorious black smokestacks. Many great cities of the world, including London and New York, are still partially stained by the years when coal was burnt in open domestic fireplaces. And, of course, coal in its raw form is black, is dusty, and prone to spontaneous combustion if not stored properly.

Scientific advancements, along with human rights developments in our society over the past century, have led to a very different coal industry to that of the Industrial Revolution, though the industry seems to be able to do little to overcome the stigma of that era. In the developed world, occupational safety hazards and worker exploitation have been almost completely eliminated, while in the developing world great progress is being made in these regards. The replacement of unskilled labor with highly trained personnel, the use of advanced mechanics, and a greater understanding of chemistry and physics have been fundamental in improving operations.

Today, coal is burnt in state-of-the-art boilers that comprehensively remove ash dust, sulfur, nitrogen, and even rarely occurring heavy metals. Modern boilers are heavily filtered so all that remains from coal firing are water vapor and colorless, odorless, tasteless, carbon dioxide. (The extent to which CO_2 itself is a pollutant is amply covered elsewhere.) The ability to now build and operate boilers pollution free is not commonly appreciated. When my wife and I were on a tourist bus in China recently, a very vocal environmentalist proudly pointed out to his friends the wanton pollution of a Chinese coal-fired power station; he expressed his disgust at the Dickensian vision of clouds pluming from the flute. He was, in fact, pointing to the water vapor from the cooling towers.

The clean-up and rehabilitation of land surfaces following mining operations have been another factor in coal mining's historically foul reputation, with many past sites treating their surrounds with, at best, disregard. In many areas of the coal mining world (e.g., the United States, the United Kingdom, and Australia) where I spent parts of my youth, abandoned open pits and washery waste heaps still litter the landscape after more than half a century of neglect. In Indonesia, the modern regulated mines operating to high rehabilitation standards leave the mining area in as good a condition, and in some cases where trees have been planted, arguably better than prior to mining. This is certainly in stark contrast to the illegal operators who scar the landscape, destroy the jungle, and leave open pits that fill with water, which turns to acid poison. Every year Indonesian coal funnel millions of dollars into education, health, and animal welfare programs to aid in the responsible development of the communities they are a part of.

Indonesian mines are almost exclusively open-pit, truck-and-shovel operations. When factors are combined, draglines and highly mechanized bucket wheel and conveyors systems for overburden removal cannot be deployed. However, while the operating costs of truck and shovel operations are quite high (in the order of US $2.50–3.00 per bank cubic meter of overburden removed, replaced, and rehabilitated), the capital cost is relatively low at about US$2 million per 1 million bank cubic meters per annum of installed overburden removal capacity. We at Bumi Resources owe our livelihood to coal mining. As do more than 100,000 Indonesians who depend on the cash flows from our operations to directly and indirectly support themselves and their dependants. Mine workers employed by listed mining companies typically earn double to treble the incomes of their compatriot peers working in comparative industries, regardless of the country or its general level of development. Huge royalties and taxes paid by Bumi and other mining companies to the Indonesian Government also provide vital stimulus to the economic well-being and stability of the nation. Countries like Indonesia are perfectly poised to be at the forefront of coal's resurrection.

In the absence of extreme levels of carbon tax, electricity producers on the coast of energy-poor countries (on the arc from North Asia through South East Asia to India) will choose coal as the most commercially attractive large-scale fuel source for the next one and probably two decades. Books such as *The Renaissance of Steam Coal* by Lars Schernikau are vitally important in presenting the public with unbiased quality information, assisting in rational debate, and hopefully avoiding counterproductive public policy. It is only through the availability of information and the de-cloaking of the coal industry that the resource coal, and the marketable energy it produces, can be judged fairly as a modern solution to our global consumption requirements. With his years of market experience and as a long-standing coal business owner, Lars is in a unique position to help us better understand this coal market.

Director PT Bumi
Bumi Resources, Jakarta
Indonesia

Kenneth P. Farrell

Acknowledgements

I would like to thank my wife and my family for their continued support. I would also like to thank Professor Dr. Georg Erdmann, Professor Dr. Christian von Hirschhausen, Dr. Wolfgang Ritschel, Siegbert Domula, Stefan Endres, and Martin Seitlinger, all participants in the 2008 online coal market survey, and all other friends and business partners for their input and comments. I would like to thank HMS Bergbau AG in Berlin and Bumi Resources in Jakarta for their support and sponsorship of this research and book project.

Contents

List of Figures

List of Tables

Abbreviations and Definitions

af	Ash-free
AFT	Ash fusion temperature
AMD	Acid mine drainage. A metal-rich water, resulting during mining from a chemical reaction between water and rocks containing sulfur-bearing minerals
API2	Physical coal price index published each week by McCloskey and Argus Media. API2 is the price for 1 metric ton of coal (6,000 kcal/kg net as received, less than 1% sulfur as received) delivered CIF Europe (ARA = Amsterdam, Rotterdam, Antwerp) in Capesize vessels (~150,000 tons)
API4	Physical coal price index published each week by McCloskey and Argus Media. API4 is the price for 1 metric ton of coal (6,000 kcal/kg net as received, less than 1% sulfur as received) delivered FOB Richards Bay, South Africa
ASTM	American Society for Testing and Materials
ASX	Australian Securities Exchange
BCG	The Boston Consulting Group, international strategy consulting firm
BEE	Black Economic Empowerment or Black Economic Empowered (South Africa)
BtC	Biomass to Coal, gasification of biomass to coal products
BtL	Biomass to Liquid, liquefaction of biomass to fuel products
Btu	British thermal unit, a traditional unit of energy (1 Btu = ~1.06 KJ)
CAGR	Compound annual growth rate
CCGT	Combined cycle gas turbine
CCOW	Coal contracts of work, Indonesia
CCS	Carbon Capture and Storage
CDS	Clean dark spread defined as base load electricity price minus coal price minus price of emission rights
CHP	Combined heat and power
CHPP	Combined heat and power plant
CIF	Price cost insurance freight (Definition as per Incoterms 2000)

CtL	Coal to Liquid, liquefaction of coal to fuel products
CV	Calorific value
daf	Dry ash-free
DIW	Deutsches Institut für Wirtschaftsforschung (German Institute for Economic Research)
DS or Dark spread	Dark spread, defined as baseload electricity price minus coal price
EEX	European Energy Exchange in Leipzig, Germany
FC	Fixed carbon
FOB	Price free on board (Definition as per Incoterms 2000)
gad	Gross air-dried
gar	Gross as-received
GCV	Gross calorific value
GDP	Gross domestic product
GHG	Greenhouse gas
Gtoe	Gigatons of oil equivalent (the amount of energy released by burning one gigaton of crude oil)
Hard coal	Hard coal is defined as the sum of steam coal and coking coal
HGI	Hardgrove index
HHV	Higher heating value
ICMA	Indonesian Coal Mining Association
IGCC	Integrated gasification combined cycle
IPO	Initial public offering
JSE	Johannesburg Stock Exchange
LHV	Lower heating value
LNG	Liquefied natural gas
MIT	Massachusetts Institute of Technology
NAR	Net as-received
Nash equilibrium	In Game theory, Nash equilibrium is a solution concept of a game involving two or more players, in which each player is assumed to know the equilibrium strategies of the other players, and no player has anything to gain by changing only his or her own strategy unilaterally
NCV	Net calorific value
OECD	Organization for Economic Cooperation and Development
OTC	Over the counter
oxid. Atm.	Oxidizing atmosphere, relevant for ash fusion temperatures
PNG	Pipeline natural gas
PCI	Pulverized coal injection
RBCT	Richards Bay Coal Terminal. The world's largest coal export terminal located in Richards Bay, South Africa
red. Atm.	Reducing atmosphere, relevant for ash fusion temperatures
Remaining potential	Number of years that the coal in theory will last when taking current annual production, reserves, and resources into account

Reserves	Proven and recoverable deposits of coal considering today's technology
Resources	Overall coal resources, also referred to as 'in-situ coal'. Resources include the known coal deposits that are currently not economical or technically recoverable
ROM	Run-of-mine coal; coal that comes directly out of the mines before it has been crushed, screened, or otherwise treated
Spark spread	Spark spread is the theoretical gross margin of a gas-fired power plant from selling a unit of electricity
Steam coal	For the purpose of this study and in line with international practice I classify anthracite, bituminous, and the majority of subbituminous coals as steam coal. Steam coal excludes coking coal and lignite
T&D	Transmission and distribution
tce	Tons of coal equivalent, assumes coal with a calorific value of 7,000 kcal/kg net as-received (SKE = Steinkohleeinheit in German)
toe	Tons of oil equivalent
TSR	Total Shareholder Return, a measure to determine profitability for an investor that includes share price and dividends
VDKI	Verein der Kohleimporteure (German Coal Importers Association)
VM	Volatile matter
WW2	Second World War
WWF	World Wildlife Fund

Chapter 1
Executive Summary

This chapter summarizes the key points of the book. Chapters 2 (starting on page 17), 3, 4, 5, 6, and 7 (starting on page 147) form the core of the book. The appendices start on page 169.

1.1 Sources of Coal – Synopsis of Coal's Significance as a Resource

Coal is now the single most important source for the generation of electricity (with a 40% share), followed by gas (20%), nuclear (15%), and oil or petroleum (7%). While we in the West are used to switching on the light in the morning we may forget that about 25% of the global population, or 1.6 billion people, still live without access to electricity. By 2030, the world's population will have reached over 7.5 billion people. With increased electrification we can expect about 1.7 billion new power customers within the next 20 years. In addition, existing power customers will increase their electricity demand to catch up with the OECD average. For instance, the Chinese still only consume 2.2 MWh per capita per annum compared to Germans who consume 6.4 MWh per capita per annum.

The 1990s were a time of false energy security, following the fall of the Berlin Wall and the victory of the international coalition in the Gulf War. This led to a lack of investment in energy raw materials. The beginning of the new millennium saw a 180-degree shift. The threat of terrorism increased and prices for raw materials skyrocketed, largely driven by the new Chinese boom. Today we are seeing the world enter a new global downturn but we can expect that energy raw materials and coal will remain on the political and economic agenda.

The objective of this book is to analyze the global seaborne steam coal trade market and find answers to the following key questions: Can coal fill the energy gap between the fading *Oil Age* and the future *Solar Age*? What relative coal prices can we expect? What differentiates markets with increasing marginal cost curves such as commodity markets from constant marginal cost markets? and How will consolidation affect the coal market?

L. Schernikau, *Economics of the International Coal Trade*,
DOI 10.1007/978-90-481-9240-3_1, © Springer Science+Business Media B.V. 2010

1.1.1 Coal Basics

Coal accounts for 25% of the global primary energy and 40% of the global electricity production. Coal is 'very old' biomass. It was generated in the Carboniferous era, starting about 360–290 million years ago, through biochemical and geochemical processes. Peat develops under airtight conditions and then coal products develop from peat through large amounts of heat and pressure exerted over millions of years. Diamonds are the natural progression from coal, developing after 1–3 billion years through further heat and pressure.

The key characteristics of coal are its calorific value (the energy content per kilogram of coal often expressed in kcal/kg), its moisture content, its ash content, its volatile content, and its sulfur content. In addition to these five key values, there are up to 40 different chemical values that determine coal quality. Coal is not a homogeneous product and usually differs substantially from mine to mine. However, since logistics costs account for 80–90% of the delivered price of coal, the calorific value is the key economic variable as, ultimately, power plants buy only chemical energy content that they can transform to electrical energy.

1.1.2 Production, Reserves, and Resources

In 2006, the world produced about 5.4 billion tons of hard coal; of this, 782 million tons was exported by sea in the form of 595 million tons of steam coal – the subject of this book – and 187 million tons of coking coal. Known coal reserves (proven and recoverable deposits) amount to 736 billion tons, or a 'life expectancy' of 137 years. When known resources (overall resources, also referred to as in situ *coal*) are added, the life expectancy of hard coal extends to almost 1,800 years.

Coal reserves and resources, as well as coal production, are well distributed across the globe; China, the United States, Russia, Southern Africa, and Australia (in that order) account for the majority of reserves. The five key exporting countries of seaborne steam coal are Indonesia (26% of total), Australia (20%), South Africa (12%), Russia (12%), and Colombia/Venezuela (10%). Other exporters include the so-called *swing* or *fringe* suppliers China and the United States. Australia will most likely be able to significantly extend its export capacity to cover future coal demand. It is estimated to account for almost 60% of all planned and agreed steam coal export mining expansion in the entire world, totaling 292 million tons by 2011. Russia will export more out of its Far Eastern ports, while it may be hampered by growing domestic coal demand. The key concern for most coal-exporting countries is logistics. Inland transportation capacities, such as railway capacities in South Africa and Russia and export port capacities in South Africa or Australia, are bottlenecks that can only be overcome with efficient investment programs. However, where inland logistics are government-controlled, as they are in many countries, inefficiencies and mismanagement often hamper coal export volumes.

China plays a special role as a fringe supplier in the global seaborne steam coal market. It produces about 50% of global coal volumes and historically has been an important supplier to Japan, South Korea, and Taiwan. More recently, China has become a net importer, but uncertainty remains about future coal export volumes that may disrupt the global market. Consumers need to prepare for this 'Chinese uncertainty' by focusing more on long-term purchasing, while producers need to keep their marginal cost low in order to be competitive in times of Chinese exports. Traders should keep their eyes on China for potential imports and exports and should maintain a presence in China.

The fringe suppliers, such as China and the United States, have an economic interest in the main supplying countries capturing monopoly or scarcity rents by exercising market power. Exercising market power can increase the market share for competitive fringe suppliers.

Companies can make optimal use of their resources by employing the economic theory first explored by Harold Hotelling (1931), deciding either to leave the reserve unmined until market prices rise or to start mining the reserve now and invest the resulting profits in the capital market. Hotelling defined the scarcity rent or the premiums on the marginal cost of reserve production. The scarcity rent corresponds to the economic value of the reserve in the ground. In efficient markets, one pays this scarcity rent when acquiring a natural resource.

1.1.3 Competitive Supply

The top five coal-mining companies – BHP, Xstrata, Anglo American, Adaro, and Rio Tinto – account for almost one-third of global hard coal exports. In fact, the five major steam coal exporters – BHP, Anglo American, Xstrata/Glencore, Rio Tinto, and Drummond – account either directly or indirectly through marketing agreements for over 80% of Colombian and South African exports, for over 60% of Australian exports, and for about 40% of Russian and Indonesian exports. I expect supply consolidation to continue. However, coal production in countries such as Russia and Indonesia will remain more fragmented than in others for the foreseeable future, driven by the political and legal environment.

1.1.4 Coal Mining

Based on my research and development of the 2006 marginal cost curve for global steam coal supply to the international markets, the average cost breakdown for coal mining is as follows:

- average marginal FOB costs total US $29/ton (US $26/ton for surface mining and US $38/ton for underground mining);
- pure mining costs account for almost 40% of FOB costs, while inland transportation and loading port transshipment account for the remaining 60%; and

- average global pure mining costs for surface mines total US $10/ton in 2006, compared to US $15/ton for underground mining.

Coal mining operational costs mainly comprise fuel (30–40% of all surface mining costs) and maintenance/repair (also 30–40% of all surface mining costs). Thus, investments in coal operations can significantly drive down variable costs. Interestingly, coal is projected to receive only 2% of global cumulative investments in energy between 2001 and 2030. Oil and gas are projected to each receive 19%, or almost 10 times as much as coal, even though coal accounts for 40% of the global electricity and 25% of primary energy generation. Two factors result from this 'underinvestment:' (1) coal carries less investment risk as it only requires US $3.4 per ton of coal equivalent (tce) of production growth (vs. US $19.6/tce of production growth for gas) and (2) investments in coal will increase in absolute and relative terms and therefore exert price pressure on coal assets. Investment costs will influence the market price, especially in times of scarcity when new suppliers want to earn a reasonable return on their investment before entering the market. Christoph Kopal (2007) estimates that investment costs average about US $45–50 for each new annual production ton of coal-mining capacity. In remote locations this figure can rise to US $160 including infrastructure investments, while in some smaller mines in Colombia, South Africa, or Indonesia the costs can be as low as US $3–10 per annual production ton.

1.1.5 Environmental Issues and Safety of Coal Production

The production of coal strains the environment, just as most other industrial processes do. The key environmental effects of coal production include (a) emissions from fuel-consuming equipment for mining and transportation; (b) land disturbance and mine subsistence; (c) water, dust, and noise pollution; and (d) methane emissions. Modern mining operations minimize all of the above but, with fuel comprising 30–40% of total surface mining costs, it is clear that a significant amount of oil-related emissions is caused by coal mining.

The safety of coal production has improved continuously in recent decades. Modern mines rarely report safety problems. The reported accidents most often occur in older mines in Eastern Europe and Asia. Most countries today curb the use of unskilled labor in underground mining operations, which was a major cause of accidents.

1.2 Use of Coal – Power Generation and More

1.2.1 Global Demand

Of all hard coal mined globally, about 74% is used for electricity generation and 13% each for the steel industry and other industries (mainly cement and heat production). Of all steam coal produced, about 85% is consumed in power plants. I estimate that as much as 90% of all seaborne traded steam coal is consumed in the generation of electricity.

While about 40% of global electricity is generated using coal, China and India rely to 79 and 70%, respectively, on the use of coal for electricity generation. Considering that China and India will have the largest impact on electricity demand growth in the decades to come, it is not surprising that coal is expected to increase its electricity share to 46% by the year 2030. Thus, the renaissance of steam coal is about to begin; the New Coal Age is ready to fill the gap between the Oil Age and the future Solar Age.

The world's most important steam coal importing countries are Japan with 19% or 114 million tons of global steam coal imports in 2006, South Korea with 10%, Taiwan with 9%, the United Kingdom with 6%, and Germany with 5%. Japan, especially, has historically been able to base its industrial growth on good relationships with raw-material-rich countries such as Australia. I conclude that the overall demand for seaborne steam coal will continue to increase at a rate higher than the basic demand growth for coal-based electricity. This growth will mainly be driven by Asian countries, namely India and China; however, Europe's demand is also likely to remain strong as local coal production is gradually being replaced with imports.

1.2.2 Power Markets

Electricity is of such importance for any country's economy that tight regulation is required to ensure reliable and stable supply of electricity to industry and households. Only governments can enforce overcapacity to handle 'extraordinary' peak demand. Private companies would not be economically incentivized to hold such overcapacity in generation or transmission assets. Nevertheless, liberalization, deregulation, and unbundling have led to a reduction in monopoly structures in the western world and parts of Asia. Today, in most European countries we find an oligopolistic competitive structure in power generation.

1.2.3 Power Generation

Power plants utilize coal for the generation of electricity. In the plant's boiler the chemical energy contained in the coal is converted into thermal energy in the form of steam. The thermal energy of the steam is then converted into mechanical or kinetic energy in steam turbines before the resulting mechanical energy is converted into electrical energy in the generators. Not surprisingly, much energy is lost in this process. Global coal-fired power plant efficiency averages 30%. While German power plants yield an average efficiency of 42%, those in China and Russia reach only 23%. Future technology is expected to reach efficiency levels of 60%, not counting energy utilized for potential carbon capture and storage (CCS). For illustration, if

Chinese power plants were to achieve average German efficiencies, then as much as 1.1 billion tons of CO_2, or 4% of global CO_2 emissions, could be saved.

The characteristics of coal can influence power plant efficiency most during and before the conversion of chemical energy into thermal energy in the boiler. The calorific value coupled with the moisture content, ash characteristics, volatiles, and physical characteristics will influence the effectiveness of coal preparation and coal burning processes. Differences in seemingly similar coal qualities can account for efficiency increases or decreases of as much as US $5–10 per ton when calculated back to the coal price. This is why it is important that technical personnel in the power plant and commercial personnel in the procurement department need to work together closely to optimize the overall profitability of a coal generator. It is highly recommended that coal power plants are able to burn most kinds of coal, also that of lower calorific value, in order to allow for flexibility in procurement. However, coal quality can make a large economic difference for the power plant during the combustion process.

1.2.4 Environment Issues Associated with the Use of Coal

The key environmental concern with coal burning is the emission of CO_2, and rightly so. In 2006, 40% of the 26 billion tons of CO_2 released into the atmosphere by human activity was caused by the combustion of coal products. Around 74% of all CO_2 emissions from power generation was caused by coal and only 13% by gas. Assuming standard efficiency of 38%, each standard ton of coal (6,000 kcal/kg NAR, 65% carbon content AR, 25% volatiles AR) results in 2.65 MWh of electricity and 2.36 tons of CO_2. Thus, each 1 MWh of electricity generated from such coal emits 0.89 tons of CO_2.

We can already infer that the best way to reduce greenhouse gas (GHG) emissions is to increase efficiency and to burn high CV coal with low relative carbon content. In terms of technology, the integrated gasification combined cycle process (IGCC) is the most promising for capturing CO_2. This process captures GHGs before combusting the fuel. Today, modern IGCC processes can already yield 40% power plant efficiency. However, such power plants still remain very expensive, costing up to 80% more than a similar standard coal-fired power plant.

1.2.5 Comparative Analysis of Coal Substitutes

Substitutes for coal are the traditional fuels gas and oil, uranium for nuclear energy, and alternative fuels such as hydro, wind, and others. The most relevant substitute for coal-fueled power generation is gas-fueled power generation. Gas emits about 45% less CO_2 per MWh than coal during power generation and requires lower investment costs. Gas can also be combusted more efficiently, resulting in less waste of the energy resource, and gas power plants can be switched off and on more quickly, in 5 min or less. On the other hand, coal has a larger resource base and is more widely distributed across the globe. Also, coal is not as tightly monopolized as gas, which – for export – is mainly sourced from less politically stable regions in Russia and the Middle East. The production of coal is also simpler and demands less investment. In particular, the transportation of coal is much easier and more cost-efficient than building very expensive transmission pipelines or LNG terminal systems. Coal also generates electricity at lower variable costs than gas and historically has been more profitable for utilities. Gas is mostly used to satisfy peak demand in Europe, while coal is used more for off-peak demand.

LNG already accounted for about 24% of the global cross-border gas trade in 2006. LNG is expected to be more economical than pipeline natural gas (PNG) for distances above 2,000–3,000 km. The key LNG importing country is Japan, with 39% of traded volume. The biggest exporter is Qatar, with 15% of total traded volume. The IEA predicts that LNG will account for 50% of the global cross-border gas trade by 2030.

It is generally believed that coal and gas prices have little co-integration while gas and oil are very much correlated. Since the principal application of gas (electricity) is very different from the principal application of oil (fuel for transportation), while gas and coal share the same application, this fact seems counterintuitive. From a purely market-oriented perspective, coal and gas should correlate more than gas and oil. Historically, however, gas prices have contractually been linked to oil prices and until this contractual link is broken we will continue to see this somewhat illogical correlation.

Nuclear power emits virtually no CO_2. However, there are three main risks associated with nuclear power: (a) accidents in nuclear power plants, (b) disposal of radioactive waste, and (c) abuse of nuclear fuel/terrorism. These risks are so substantial that I do not predict that nuclear power will gain significantly more in importance than it already has. France is the one exception, where 78% of electricity is currently generated using nuclear power. Other developed nations use more in the range of 25%. Globally, nuclear power accounts for 15% and oil-based power for only 7% of power generation. Oil based power will continue to decline as countries that heavily rely on oil for power, such as Indonesia, Mexico, and Morocco, are already switching from oil to coal and/or gas.

Alternative fuels such as hydro, wind, biomass, and solar can be divided into two categories: (1) the traditional alternative hydro and (2) the newer alternatives wind, biomass, solar, and others. Global nonfossil and nonnuclear power generation accounted for 18% or 3,300 TWh in 2005. Of this about 85% comes from hydro; thus less than 3% of total global power, or approximately 500 TWh, stems from wind, biomass, solar, and other sources. It is safe to assume that the majority of the 500 TWh stems from wind. The problems with wind are (1) that much of it is still heavily subsidized, (2) that it is erratic and generation cannot be predicted, and (3) that it requires large areas to generate any noticeable energy. For instance, 1,500 2-MW onshore wind turbines are required to replace a single 1,000-MW coal-fired power plant, assuming that the wind turbines work at full capacity 33% of the time (which is a very optimistic assumption for onshore wind parks).

The IEA estimates that non-hydro generation will triple to 1,500 TWh by 2030. Given the demand increase of about 16,000 TWh within the same timeframe, this means that only 6% of the power demand increase up to 2030 can be covered by non-hydro alternatives, while 94% (of which only 5% is hydro) will be covered by coal, gas, nuclear, and hydro. We can conclude that fossil and nuclear power will contribute above average to the increase in electricity generation up to 2030, and in the next two decades, alternative fuels will not even be able to maintain their current share in the electricity mix (see Fig. 4.14 on page 102). This, however, does not mean investments in such alternatives should be stopped; we just need to face the reality and deal technologically with fossil fuels, foremost with coal.

It can be concluded that coal is best suited to covering additional electricity demand, triggered by a population and electrification increase in the next two to three decades. However, the key environmental problem of coal, namely high CO_2 emissions, needs to be resolved. Long-term non-hydro and solar-based alternatives are the only way to meet the planet's energy needs. However, the development of economically feasible alternatives will take longer than the scope of this study allows, and the problem is more likely to be solved toward the end of the century.

1.3 Global Steam Coal Market and Supply Curve

1.3.1 Geopolitical and Policy Environment

Most if not all countries in North America and Europe still lack a coherent energy policy. National energy policies are about optimizing a nation's development and wealth. As a result, resource-rich nations will support policies that tend to increase international raw material prices. It is the task of international organizations to find a globally consistent way of exploiting the planet's natural resources and to support environmentally friendly processes and regulations. The Kyoto Protocol was a good first step but has caused a trading scheme in Europe that in fact may increase, rather than reduce, global CO_2 emissions. For illustration please consider that a European-based trading scheme will penalize coal and favor CO_2-friendlier natural gas, which seems sensible at first glance. As coal is penalized, the price of coal will be forced down relative to gas. Outside of Europe, supported by European policy, coal is now cheaper than gas. Thus, in countries with lower process efficiencies such as China or India – countries that consume far more coal – coal has suddenly become even more attractive. In the end, the Europeans will have moved coal-fired electricity away from high efficiencies in Europe to countries with lower efficiencies that consume larger volumes and grow faster.

I suggest that policies need to be adjusted and that policymakers should learn more about coal and energy resource economics. Energy efficiency needs to be encouraged, but any measure has to be consistent with environmental constraints and sound economics. Kyoto and emissions trading need to be reconsidered; it is unfortunately not economical or environmentally friendly to start such schemes without including large CO_2-generating and energy-hungry nations such as the United States, China, and India. The price gap between coal and gas needs to be closed in order to encourage industry to burn more gas than coal. However, gas is limited, so it is advisable to invest in optimizing the use of coal and in clean coal technologies.

1.3.2 Global Seaborne Steam Coal Trade Market

The coal trade has traditionally been divided into the Atlantic and Pacific markets. In the Atlantic market, Europe is supplied mainly by South Africa, Russia, and Colombia/Venezuela. In the Pacific market, Japan, Taiwan, and South Korea are mainly supplied by Australia, Indonesia, and, increasingly, Russia. The two markets interact closely, especially in times of lower freight rates. It has been shown that the law of one price also applies to the global coal market.

Market power has shifted from the consumers to the producers. Producers are more consolidated and are being offered supply alternatives due to increasing domestic coal demand in the coal-exporting countries such as Russia, South Africa, and Indonesia. Physical coal traders serve a number of functions in the coal market, including but not limited to (a) acting as a physical buffer, (b) financing cargoes, (c) arranging freight and logistics, (d) acting as outsourced purchasing or sales departments, and (e) mitigating credit risk. Trading, by nature, is a much more volatile business than coal production or electricity generation.

Coal is shipped in bulk carriers across the ocean. Shipping is both cost- and time-intensive. Bulk carriers fall into three categories: Capesize vessels, Panamax vessels, and handysize vessels. Almost 60% of sea bulk cargoes comprise steel-related cargoes such as iron ore and coking coal. The two other major bulk products that influence bulk freight are steam coal and grain. Freight prices reached unprecedented price levels in summer 2008 when some Capesize vessels sold for over US $200,000/day. In fall 2008, freight prices dropped to less than one-tenth of that level and have reached prices below US $10,000/day, sometimes below the marginal cost of operating a vessel. Thus, volatility in the freight business has greatly surpassed volatility in the commodity business. As a result, sea freight makes up varying percentages of the delivered coal price, ranging from 10% to over 50% of the delivered price.

Joskow (1987) wrote the first widely known scientific paper about coal contract terms. His results support the view that buyers and sellers make longer commitments to the terms of future trade at the contract execution stage and rely less on repeated bargaining, when relationship-specific investments are more important. Today, contract term behavior has drastically changed. Many consumers buy more spot and many contracts are index-linked, based on either the API2 or the API4 indices. The development of coal derivatives took off after 2000 and has not only changed contract term behavior. Today, companies can buy and sell coal forward either to hedge or to speculate. In 2007 the paper physical multiple reached 10x (still far away from oil's 4,000x). There are a number of problems with the current coal derivatives: (a) they are still mostly traded OTC and not at an exchange that manages default risk and (b) the coal price index is created by calling around the market rather than through standard and real offers/bids. Exchange-settled coal contracts (traded on ICE or ASX) are quickly catching on though, fueled by the 2008 financial crisis. Paper hedging is not suitable for individual or smaller contracts since paper does not recognize force majeure or delivery delays.

1.3.3 Pricing and Capacity Utilization

Coal is a natural product and not homogeneous in quality, though successful attempts have been made to commoditize at least part of the business (see globalCOAL), especially South African coal. Also, each supply region has vastly different market conditions, often driven by infrastructure, quality, and location, and many destination ports are monopolized or controlled by one or two consumers. Given that not all ports can accept all vessel sizes, this leaves room for price discrimination. To simplify coal pricing and to start modeling the coal market, I have identified five key determinants: (1) marginal FOB costs and elasticity of supply; (2) export mine capacity; (3) demand growth; (4) emissions prices; and (5) sea transportation. Marginal FOB costs are discussed in the next section. I shall point out here only that export mine capacity utilization has more recently consistently topped 90%, resulting in scarcity. While this may ease for some time during periods of crisis, I have shown that generally we can expect coal scarcity to continue supporting higher prices that are above perfectly competitive price levels where price equals marginal cost.

1.3.4 Variable Cost Analysis – Real Global FOB Costs

I have constructed an FOB marginal cost supply curve for the global seaborne steam coal supply market for the years 2005 and 2006. I was able to draw from over 100 personal interviews and a wide range of experience, including but not limited to trading coal products of all coal origins other than Australia, and analyzing, investing in, or partnering in coal mining projects in South Africa, Colombia, Russia, and Indonesia. The result is a three-dimensional matrix of FOB variable costs, which I analyzed for 2005 and 2006, detailing (1) supply region, (2) surface vs. underground mining, and (3) costs (pure mining costs, inland transportation costs, and port transshipment costs). Poland and Russia turned out to be global marginal suppliers determining the price in perfect competition. Indonesia and Australia are the lowest cost producers. Average 2006 FOB costs were US $29/ton (2005: US $28/ton), with the marginal supplier Poland reaching

US \$61/ton. The average of all FOB marginal costs for underground mining (less than one-third of all export mines) is US \$38/ton compared to US \$26/ton for surface mines. Thus, underground mining is close to 50% more expensive than surface mining on an FOB basis. I have also shown that the marginal cost for coal production is not constant, but in fact increasing, and thus variable. This phenomenon differs to normal manufactured goods, but is experienced with other raw materials. As shown in the previous section, the coal market is not perfectly competitive. Thus, game theory and Cournot may be used in addition to standard perfect competition to scientifically describe the market.

1.3.5 WorldCoal: GAMS-Programmed Coal Market Model Including Sea Freight

WorldCoal, a nonlinear model to qualitatively analyze the global steam coal trade market, including sea freight, was developed under my supervision by a research team at the Technical University of Berlin. The model enables real-time analysis of the current market situation, which was carried out for the year 2006. Using this model, it is possible to analyze certain likely future scenarios, such as production capacity or logistical constraint scenarios. The model uses GAMS programming to optimize distribution of the 2006 coal supply in such a way that the total cost of the market is minimized.

A number of restrictions were defined and the global market was simplified using data from the real FOB marginal cost analysis previously discussed. The world was divided into eight supply regions and nine demand regions, and a supply–cost curve was developed for each supply region. Freight was then approximated in the WorldCoal model by allocating one loading port per supply region and one discharging port per demand region.

The model's results for the reference year 2006 were surprisingly close to the real market. Market prices differed on average only by about 6%, despite the model's many simplifications, including but not limited to the one-quality assumption. This indicates that the 2006 coal market operated close to the theoretical market equilibrium in a perfect competitive world. The 2015 scenario restrictions were such that supply becomes even more concentrated, with many demand regions having only one supplier. In the real coal world, this would not happen. It is unlikely that one demand region will take the strategic risk of depending on one single supply region for any longer period of time, even if it would be cost efficient. Such strategic thoughts could not be modeled into WorldCoal, however.

WorldCoal models an entire year. Thus, any intra-year volatility cannot be described, and in fact it wasn't meant to be. When modeling a global market, the point is to get an understanding of what will happen when certain scenarios occur. They are also useful for predicting long-term market trends. As such, I believe that WorldCoal is fulfilling its purpose for the global coal market. However, WorldCoal has limitations that future researchers will hopefully reduce. As with any model, it has to remain manageable and thus will always require simplifications.

1.4 Industrial Structure of Supply Market: Game Theory and Cournot Competition

In this study, Cournot's constant marginal cost assumption was relaxed to include increasing marginal cost and to discuss the impact on price and profits. Despite the simplifying assumptions that the competing companies are symmetrical and their cost function is quadratic, the following conclusions can be made about Cournot competition, confirming some market intuition:

1. Markets with increasing marginal costs, such as natural resource markets, demand a higher price premium under Cournot competition than markets with constant marginal cost, such as standard manufacturing-based markets.
2. A shortage situation in a natural resource market with steeper marginal cost curves results in a higher price premium in a Cournot model.
3. Symmetrical players with increasing marginal cost curves in a market (i.e., a scarce market) will earn less profit than players in a market with constant marginal cost curves (i.e., without scarcity).
4. The more the players participate in an increasing marginal cost Cournot market, the lower the price and profits, and the same applies to constant marginal cost Cournot markets.

For the coal market this means that a scarce market with steep marginal cost curves results in higher prices and lower profits for the marginal producer. On the other hand, non-marginal producers are interested in scarce markets as scarcity increases price and, hence, profits. More realistically, let's take as our example a supply market with a small number of fairly symmetrical coal exporters (e.g., Russia) and a demand market that is supplied almost exclusively by these exporters (e.g., Finland). Here one can apply well the new increasing marginal cost Cournot model. The exporters are able to translate their market power into increased profits. The Russians would not like a true scarce-supply market; in fact, they would do everything possible to keep the market affluent and supply at the equilibrium price. The scarcer their market becomes (or the steeper their marginal cost curve becomes), the lesser the profits they will be able to extract based on their market power.

From the above it can be concluded that market players have an incentive to continuously invest in production to make the marginal cost curve flatter. At the same time, investments will increase barriers to entry and thus play a part in protecting a player's market position. It was also proven that players in an increasing marginal cost Cournot market are interested in a reduced number of market participants, thus resulting in consolidation. This has already been seen in the coal market, in which the top five coal exporters in 2006 (BHP, Anglo, Xstrata/Glencore, Rio Tinto, and Drummond) control Australia, South Africa, and Colombia with about 67–86% of exports. Even in Russia and Indonesia these five players directly or indirectly account for almost 40% of exports.

1.5 Conclusions, Implications, and the Future of Coal

The coal trade is expected to continue growing above coal demand growth rates as globalization continues to translate into increased trade. The resulting higher trade volumes will continue to professionalize the coal market, attracting new highly-educated talent and increasing transparency. Demand increase will be driven by the Pacific market, mostly India and China. The five main export regions – Australia, Indonesia, South Africa, Russia, and Colombia/Venezuela – will not equally benefit from increased export volumes. Logistical constraints and domestic coal demand will keep a lid on exports, especially out of South Africa, Indonesia, and, to some extent, Russia. China is the wild card for global coal supply since it plays a special role as a fringe supplier. Historically, China has been an important supplier to Japan, South Korea, and Taiwan. More recently, it has turned into a net importer, but uncertainty remains about future coal export volumes that may disrupt the global market.

Consolidation in the supply market makes economic sense for producers' shareholders as it leads to higher EBITDA and TSR. However, from a macro-perspective, perfectly competitive markets are desirable. Thus we have two trends, one driven by economics and the other driven by politics. Public policy will develop to manage coal supply consolidation more closely. As much as policy will allow, we will continue to see more and larger merger attempts in the coal supply arena. Market participants will invest more in logistics and upstream assets as coal remains a scarce resource market. This is also supported by the fact that pure mining costs account for only 10–20% of the delivered coal prices; the remainder are logistics-related costs.

New market participants, including banks, will appear on the physical coal market. Default risk will increase credit risk requirements and make trading more expensive. Trade risks will increase and the same coal cargo will be traded through more and more companies. I expect that exchange-based coal trading will develop in the coming decades but will be limited to larger standardized coal volumes. This will include the standard RB-ARA and NEWC-ARA route. I predict that at least 50% of coal volume will continue to be traded OTC in the long term, driven by coal quality and fragmented trading routes. Coal derivatives will continue to grow. The recent financial crisis will curb growth but in the long term new financial traders will continue to influence the market and strive for more index-based pricing.

1.5.1 Future Steam Coal Price Trends

Current trends point toward higher steam coal prices driven by (1) increasing electricity demand; (2) increasing share of coal in electricity generation; (3) increased domestic coal demand in major coal-exporting countries; (4) FOB cost increases due to labor, fuel, machinery, and equipment; (5) coal asset price increases because relative coal investments will slowly catch up with oil and gas investments; and

(6) producer consolidation. For global commodities as well as for coal, 2008 was an extraordinary year, with CIF prices peaking at US $210/million ton in July 2008. In the longer term, coal prices will continue to rise, starting again in 2010. Coal price volatility will also increase, but should remain below gas and oil price volatility.

I predict that coal prices will be above marginal cost of production in the long run because producers have enough market power to keep average coal prices above perfectly competitive market price levels. Short-term price fluctuations may, however, cause coal prices to fall below marginal FOB costs for brief periods of time. I also expect coal prices to slowly catch up with gas prices because I believe that the basic economic principle of making a CO_2-friendly fuel less expensive than a non-CO_2-friendly fuel such as coal will win support from policymakers.

1.5.2 Future Sources of Energy

What role will coal play in the global power mix? The short answer is coal will become more important than it already is for primary energy, and even more so for electricity generation. The renaissance of steam coal is just about to begin.

World primary energy consumption is growing at an average annual rate of about 1.7% (CAGR) for the period 1990–2030. Electricity, on the other hand, has always grown much faster and will continue to do so at an average annual rate (2005–2030) of 2.6%. Growth is fueled by non-OECD countries, most importantly China and India. As such, electricity generation will grow at a CAGR (2005–2030) of 4% in non-OECD countries and 1.3% in OECD countries. I have shown that coal is expected to increase its share of electricity generation from 41% in 2005 to 46% by 2030. Only gas will continue to grow slightly above coal with a CAGR (2005–2030) of 3.7% compared to 3.1% for coal. As expected, oil-based electricity generation will retreat. More interesting is the fact that renewable energy sources will retreat as well in relative terms, at least until 2030. Renewable electricity generation simply cannot keep up with increased demand to bridge the gap until the technology is available to exploit solar energy in a more efficient and economic way.

Richard’s Bay Coal Terminal, South Africa

Coal truck in Port of Karachi, Pakistan

Open cut coal mine in South Africa

Rainy season in Indonesia at a coal stock pile

Chapter 2
Introduction

2.1 Prelude

The world today depends on the fossil fuels oil, coal, and gas (in that order of importance) for over 80% for its primary energy. From the time early humans tamed fire, wood or biomass became their primary energy sources. Coal took over the leading role from biomass during the Industrial Revolution and accounted for over 60% of the world's primary energy by the early 1900s. The current age is often referred to as the Oil Age, which is somewhat appropriate considering that about 35% of the world's primary energy still comes from oil. However, today about 25% of the world's primary energy and more than 40% of the world's electricity comes from coal. In addition, about 66% of the world's steel is produced using coal (IEA-Statistics 2005; VDKI 2006).

The World Coal Institute projects that coal will again become the primary source of energy in the future (see Fig. 2.1). Having worked for over 6 years in the international coal industry and having studied the economics of the steam coal market for over 3 years, I agree and, therefore, this study is also entitled, 'The Renaissance of Steam Coal.'

The world's appetite for energy is still far from being met. In 2002, 1.6 billion people or almost 25% of the world's 6+ billion inhabitants were still without access to electricity. Of the remainder, 2.4 billion people were dependent on primitive or erratic electricity supply. In 2030, the world's population is expected to have reached 7.5+ billion. By then, about 1.4 billion people will still lack access to electricity (VDKI 2006). Thus, it is expected that there will be about 1.7 billion new power customers in the next two decades. The international strategy consulting firm *The Boston Consulting Group* has tellingly named a series of its successful economic consumer studies 'The Next billion Consumers,' indicating the huge growth in demand that the energy industry is faced with. The growth in energy demand is primarily driven by non-OECD countries such as China and India (see Fig. 2.2).

We can expect significant improvements in productivity due to technological advances in the next 50 years. The European Commission (2006, p. 12) estimates in its reference projection that the world economy will increase fourfold in that period. At the same time, the world's energy consumption is projected to increase by only

L. Schernikau, *Economics of the International Coal Trade*,
DOI 10.1007/978-90-481-9240-3_2,

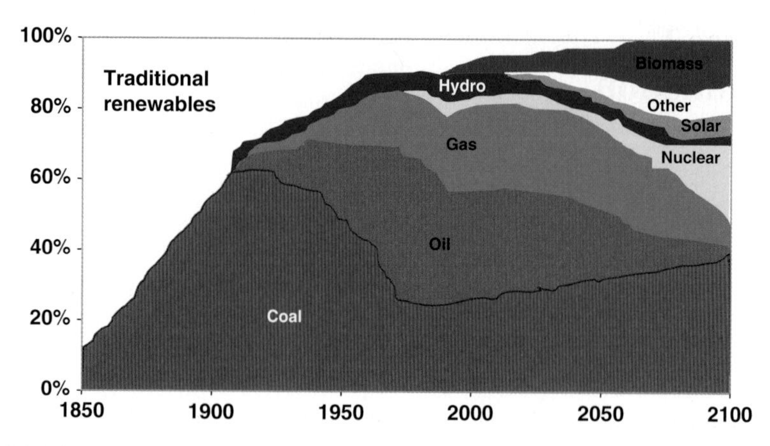

Fig. 2.1 Historic and future world primary energy mix: Scenario 2100 (Source: World Coal Institute, http://www.wci-coal.org)

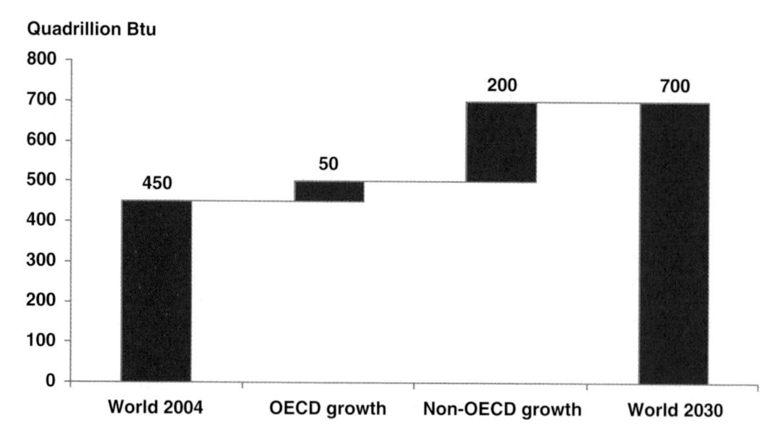

Fig. 2.2 Comparison of energy demand growth up to 2030: OECD/non-OECD (Source: EIA 2007; Author's analysis)

a factor of 2.2 from about 10 Gtoe (gigatons of oil equivalent) in 2006 to about 22 Gtoe in 2050. However, 'only' is a relative term, as this task is more than difficult enough for the human race and the environment.

In the 1990s, certain events and market phenomena created a false sense of energy security in the Western world (see Yaxley, 2006). The Berlin Wall fell, and with it a decades-old enemy system. The victory of the international coalition in the first Gulf War and further European Union expansion deepened the false sense of geopolitical security. There was also a tendency to misinterpret energy policy as an extended arm of climate policy. Politics and modern environmentalism increasingly regarded coal and nuclear energy as a scapegoat. This attitude was coupled with an overestimation of the short- and mid-term potential of renewable energy. Overcapacity in coal, oil, and gas led to low fossil commodity prices in the late 1990s and the early 2000s, which in turn resulted in dangerous oversight of the unequal distribution of resources and the limitation of energy resources in the world. The resulting lack of investment by producers led to false expectations by consumers.

However, the new millennium also brought a set of new circumstances. Today, in the beginning of the third millennium, there is a growing need to reassess energy policy and to become aware of the importance of fossil fuels, especially coal, and our reliance on them. The threat of terrorism has increased dramatically, introducing a new type and concept of enemy for the West. Also, more political problems and rising instability in supplying countries have shaken the Western world. Oil, coal, and gas prices have skyrocketed (even allowing for the price drops in fall 2008), again raising questions about the impact of monopolistic and oligopolistic markets on the world economy, and, as a result, the role of governments and protectionism. Today, renewable energy sources are being reevaluated and their potential estimated more realistically than in the 1990s. In addition, the Chinese economic boom has affected every aspect of the world economy, including almost all commodities and logistical capacities.

As a result, commodity prices have increased sharply (see Fig. 2.3). Along with them, by summer 2008, coal prices reached unprecedented levels, tripling in price within 1 year (see Fig. 2.4) before falling again to still historically high levels. The appetite for renewed investments is demonstrated by a tenfold increase in the value of the market capitalization of the top 10 coal stocks – from US $6.6 billion on January 1, 2000, to US $69.3 billion on October 19, 2006 (see Tory 2006, p. 2).

The world seaborne steam coal market continues to be an interesting and, for the energy economy, crucial playing field. A lack of investment since 2000 (Kopal 2007), near-capacity export production, and a new breed of market players influencing prices have ensured continued if not increased price volatility.

In the decades to come, there will be no way around coal. I propose that governments and organizations spend more time and financial and human capital on

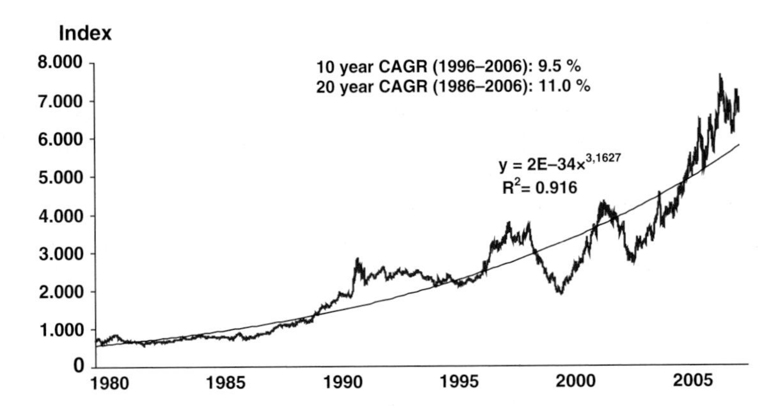

Fig. 2.3 GSCI development, 1980–2006 (Source: Author's analysis; Goldman Sachs Commodity Index, GSCI)

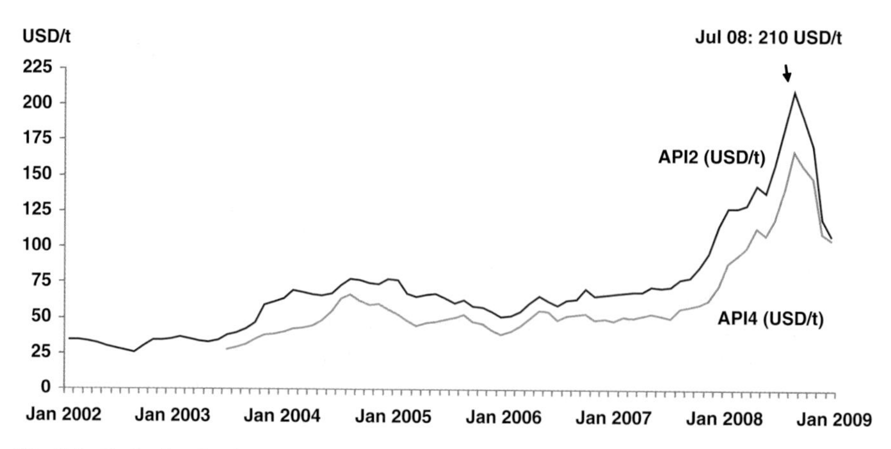

Fig. 2.4 Coal price developments, 2002–2008 (Source: Author's analysis; McCloskey Coal Price Index)

developing technologies to improve the world's power plant park and to find newer and better ways of producing and transporting fossil fuel resources rather than fighting coal (in Germany this opposition runs parallel to the fight against nuclear energy). There is no question that the world needs every megawatt-hour sourced from renewable energy. However, for the foreseeable future, renewable energy will not suffice to satisfy the world's hunger for energy. Since coal's lifetime far surpasses that of other fossil energy resources (see Fig. 4.15 on page 104) it will become increasingly important, especially once the CO_2 problem of coal utilization can be managed.

There is increasing evidence that coal will not only remain one of the key sources for our energy demand but also actually gain in importance. This renaissance of coal as a resource warrants a more scientific study of the subject. The technological aspects of coal production were studied in great detail in the last century. Much is known about underground and surface mining of coal. We also know a lot about its physical and chemical characteristics, as well as the path to further improving the efficiency of coal-fired power plants. However, much less research has so far been conducted on coal markets and the coal trade. The information available on the economics of the global coal trade is very limited. The coal market is still a very private and closed market with relatively little transparency, partly as a result of the general public antipathy toward coal. In fact, coal has only been traded on an international level since the early 1980s, a development that was sparked by the oil crises in 1973 and 1979. However, even today politics and the scientific community lack the same level of knowledge about coal as they have or can refer to in the cases of oil, gas, nuclear, and, especially, renewable energy sources (see Stanford 2008).

The need for further economic analysis of the coal market is demonstrated by an increase in recent activity to better understand the economics of coal as a resource. A variety of institutions, including the *Deutsches Institut für Wirtschaftsforschung* (DIW), as well as EURACOAL and a number of universities in Germany, the United

States, Sweden, Poland, and Japan, have recently stepped up their efforts to study the importance of coal. For instance, in 2008 Stanford University started a large research project focused on coal.

I argue that coal will fill the gap between the Oil Age and the often-referred-to 'Solar Age' of the future, where renewable energy sources will satisfy the majority of the planet's hunger for electricity in specific and energy in general. In filling this gap, coal competes head-on with other sources of energy, but coal has the major advantage of being available in a relatively free market, with supply coming from developed and developing countries alike.

The Achilles' heel of coal is the justified environmental concern. Currently, coal generates more CO_2 per MWh of electricity produced than any other fossil fuel. With 40% or 11 Gt of the total 26 Gt global CO_2 emissions stemming from coal, environmental risks demand 'clean coal' technology (IEA-CO_2 2006). For more details regarding global CO_2 emissions please refer to Appendix C. Independent of the need to step up efforts for cleaner production and use of coal, it is crucial that the world increases the speed at which renewable energy sources are being developed.

The objective of this book is to answer the following key questions:

- Can coal fill the energy gap until the full arrival of the Solar Age, especially when compared to oil, gas, and nuclear resources?
- What relative coal price levels can we expect? Will the price of coal continue to become more volatile, especially as long as we are near export production capacity of steam coal? In the long term, will coal prices remain more stable than prices for other fossil fuels, especially when considering the large reserves of coal?
- How do markets with increasing marginal cost curves (i.e., commodities, including coal) behave differently from markets with constant marginal cost (i.e., standard factory-based industries)? How will consolidation affect the coal market with an increasing marginal cost curve?

This book is also meant to contribute to (1) further professionalization, (2) greater transparency, and (3) more efficiency of the coal trading industry.

2.2 Methodology

This book focuses on the supply side of the global seaborne steam coal market. Also, the importance of coal for the international power market is explored. The scope of this study is the years 2005 and 2006. Based on the available data, research, and analyses, predictions are made 20–30 years into the future.

I use economic theory to analyze various aspects of the market. Industrial economics as well as resource depletion and price formation theories are also employed. I use standard econometrics for the analysis of quantitative data.

The empirical data used to analyze the coal market are based on primary research and qualitative data. In 2006 and 2007, I conducted over 100 interviews in 7 overseas and 9 European countries. I built a bottom-up marginal FOB cost curve for the global production of steam coal based on these interviews and based on the study of publications, research studies, and relevant coal market literature. I analyzed coal cost data for the years 2005 and 2006 and used new price data to generate long-term price trends. During August and October 2008, I conducted an Online Coal Market Survey using an internet-based online survey provider http://www.onlineumfragen.com. Of the 500 coal experts contacted for this survey, 200 responded. The questions were tested on a group of 10 test subjects in two pretests with two interviewers and were optimized over a period of 3 months.

The nonlinear equilibrium coal market model WorldCoal was programmed in GAMS for the qualitative analysis of the global steam coal market. The model is designed to analyze real-time market situations as well as scenarios about possible future developments. 2006 FOB cost data as well as 2006 freight data were used to analyze the year 2006 and to make predictions about the future.

Economic game theory was employed to analyze the Cournot competition model. The Cournot model was extended to work with increasing marginal cost rather than only constant marginal cost. The Cournot model with increasing marginal cost allows the description of the industry structure of raw material markets such as the coal market. Figure 2.5 below summarizes the structure of the dissertation that forms the basis for this book.

Chapter 3 details the sources of coal, including an analysis of coal as a resource and a regional analysis of coal reserves. Chapter 4 looks at the use of coal and its current role in power generation. Coal power plant technologies as well as environmental issues surrounding the use of coal will be explored before reaching the core of this study where – building on the previous chapters – I analyze the global steam coal market and build a bottom-up supply curve as well as the WorldCoal market model in Chapter 5 (detailed calculations supporting the WorldCoal model are contained in Appendix E). I discuss relevant game theory (Cournot model

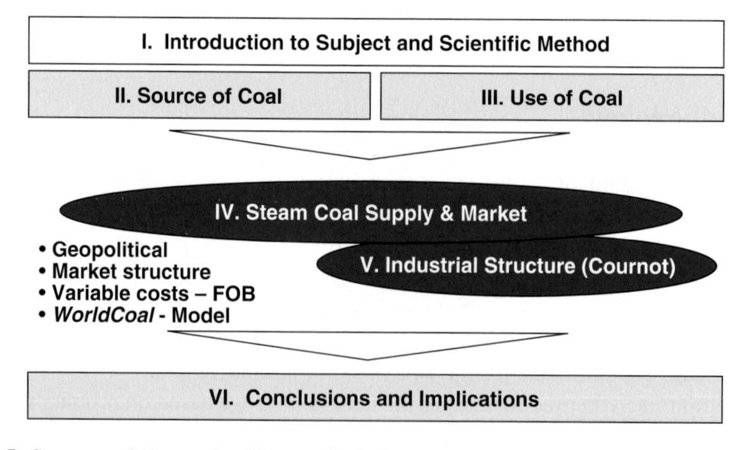

Fig. 2.5 Structure of dissertation (Source: Author)

to cover not only constant marginal cost but also increasing marginal cost) in Chapter 6. For calculations supporting Cournot analysis please see Appendix G. In Chapter 7, I summarize the book's findings and draw conclusions for the future of the international steam coal market.

2.3 Market Definition

This book analyzes the global seaborne steam coal trade. This market encompasses all steam coal (steam coal equals hard coal minus coking coal) traded by sea. It therefore does not include the coal used within one country or the coal transported across land borders, the so-called 'green border trade.'

The world produced about 5.4 billion tons of hard coal in 2006, of which 867 million tons were traded internationally. Green border trade totaled 85 million tons. The remaining 782 million tons traded by sea comprised 187 million tons of coking coal and 595 million tons of steam coal – the focus of this study (VDKI 2006) (Fig. 2.6).

Steam coal includes all bituminous and most subbituminous coals as well as anthracite. Not included is coking coal or classic lignite coal. For further details please refer to Section 3.3.

The supply for the global seaborne steam coal market is measured by the production for export. The capacity is measured by the sum of all export mine capacities. The demand is measured by seaborne trade (Kopal 2007). In this study on the demand side, I focus on the power industry's coal consumption.

This research focuses on the global market. We can do this since Li (2008) and Warell (2007) have shown that the Atlantic and Pacific coal markets are

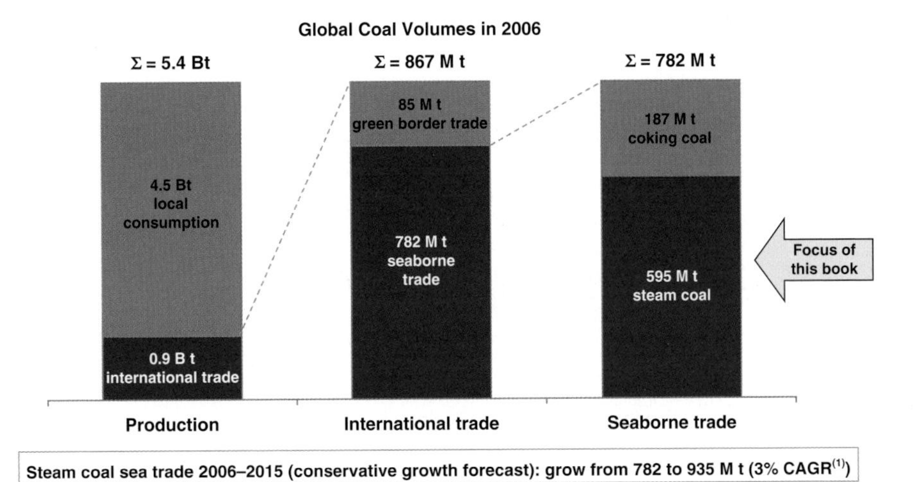

Fig. 2.6 Focus of this book: 595 million tons of internationally traded steam coal (Source: VDKI 2006; Author's analysis)

co-integrated. As a result, the traditional separation of the Atlantic and Pacific steam coal markets is fading. The steam coal market has thus become a global market and is relatively unified in terms of economics. While I will show that the coal trade market is not perfectly competitive, the *law of one price* acts as a guiding principle for defining the market. Selectively, I may refer to European data as a proxy for the global market. Data availability in this market is imperfect and available sources are not always consistent. Wherever possible I have opted to use the definitions and key figures compiled by the Coal Importers Association in Hamburg (*Verein der Kohleimporteure*, VDKI).

Chapter 3
Sources of Coal – Review of Coal as a Resource

This chapter examines the nature and significance of coal as an energy resource. It explains how coal is formed, describes its key characteristics, known reserves and resources, and how coal is distributed geographically. The author looks in detail at the major steam coal-exporting countries. He focuses particular attention on China's significance as a fringe supplier, and more recently as an importer. There follows an examination of the supply market, detailing the key players and describing a trend toward consolidation. The operational costs involved in coal mining are also examined, and the author reveals the low investment rate in coal compared to other fossil fuels. The chapter concludes with a review of the environmental and safety issues associated with coal production.

3.1 The Fundamentals of Energy Sources and Fossil Fuels

There are a number of energy sources available for human use. Table 3.1 summarizes these sources. Terrestrial sources of energy include the fossil fuels coal, oil, and gas, as well as nuclear energy based on uranium. These terrestrial sources accounted for 87% of the world's primary energy production and 82% of the world's electricity production in 2005 (see Fig. 3.1). Solar and other sources – including wind, hydro, biomass, tidal, and geothermal energy – accounted for the remainder, less than 13%. In this study we will refer to fossil fuels as primary energy sources which is, scientifically speaking, somewhat imprecise. Wolf and Scheer (2005) have pointed out that fossil fuels are merely solar energy coupled with earth matter, water, and CO_2, which have turned into biomass. In their very interesting and highly recommended book *Öl aus Sonne – Die Brennstoffformel der Erde* [Oil from the Sun – the Earth's Fuel Formula] they argue that, in the long run, the burning of fossil energy is a reversible process with solar energy required as the process energy.

When speaking about fossil energy sources, it is important to note that all carbon-based fossil energy sources are generated from biomass. Biological and physical processes in the form of heat and pressure are responsible for the generation of fossil fuels over millions of years. One could argue that even uranium is a fossil fuel; however, uranium does not develop through biological and physical processes, but rather during fusion processes in the final phase of certain stars (supernova).

L. Schernikau, *Economics of the International Coal Trade*,
DOI 10.1007/978-90-481-9240-3_3, © Springer Science+Business Media B.V. 2010

Table 3.1 Overview of sources of energy for human use

Terrestrial sources	Solar sources	Other sources
Fossil fuels	Direct solar	Tidal energy
• Hard coal	• Solar radiation	Geothermal energy
• Lignite		
• Oil		
• Gas		
Nuclear energy	Indirect solar	
• Uranium	• Wind	
	• Hydro	
	• Biomass	

Sources: Bettzuege, Marc Oliver (2007), energy: long-run sustainability, BCG industrial goods alumni meeting, Munich; Author's research.

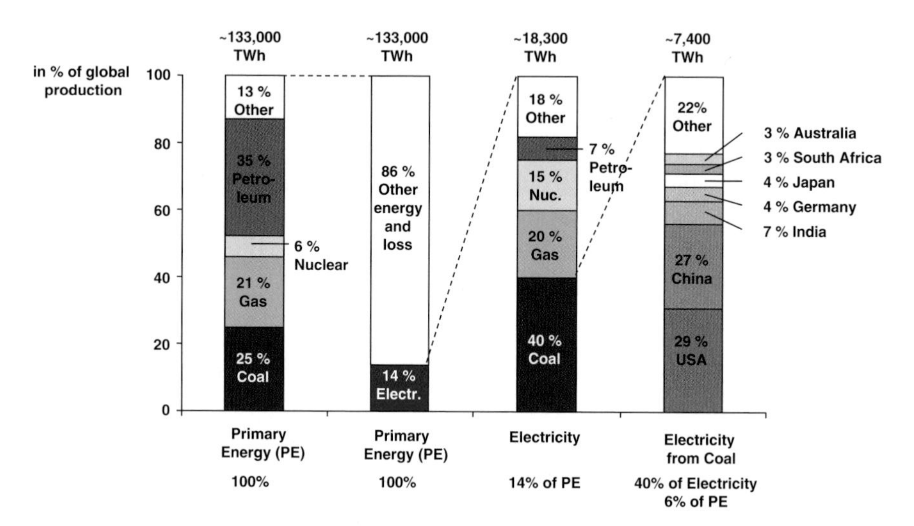

Fig. 3.1 Global shares of primary and electricity energy, 2005 (Source: IEA-Statistics 2005; Author's analysis)

Today, the majority of our electricity is generated using various industrial processes to oxidize (combust) the carbon contained in fossil fuels and biomass. In this process, molecules develop in an exothermic reaction whose bonding force (the Coulomb force) is higher than that of the original molecules (Erdmann and Zweifel 2008, pp. 15–19). The energy released in this combustion process is in the form of heat.

Formula (3.1) summarizes the amount of energy released through combustion of one kilogram of carbon content. Formula (3.2) summarizes the theoretical complete combustion of fossil hydrocarbon fuels resulting in heat plus carbon dioxide and water.

$$1\,\text{kg C} + 2.7\,\text{kg O}_2 \rightarrow 3.7\,\text{kg CO}_2 + 32.8\ \ 10^6\text{J} \qquad (3.1)$$

$$C_xH_y + \left(x + \frac{y}{4}\right) O_2 \rightarrow \text{Heat} + xCO_2 + \frac{y}{2}H_2O \tag{3.2}$$

Processes such as the steam turbine generator of a coal-fired power plant are required to generate energy in a form other than heat (i.e., mechanical or electromagnetic energy). Usually, only those fossil fuels whose energy content is economically utilized through the chemical reaction of some form of combustion are used for such energy sources. Fossil fuels are also finite. We are consuming them far faster than they can develop. This is one key fact on which environmentalists base their opposition to fossil fuels. This is also the key fact for the coal industry to consider. However, as we will see later, coal reserves and resources will far outlast those of oil and gas, and mankind continues to discover new reserves and resources. Nevertheless, the smart and economical use of our planet's fossil fuels will remain a key political and technological challenge until humans are able to satisfy their energy demand primarily from solar sources.

3.2 Process of Coal Formation

Today's hard coal resources were generated in the Carboniferous era abou 360–290 million years ago. The Carboniferous era was named for the coalification process that took place at that time. During that era the continents had not yet reached their current position. The climate was mild, and morasses, swamps, and large forests covered the land masses.

Two phases were required for the generation of coal:

1. Biochemical process: peat development under airtight conditions
2. Geochemical process: coalification through heat and pressure exerted on the peat

In the first phase, the climate, flora and fauna were important for coal to develop. Biological remains of plants such as ferns and trees could not always decay or rot fully because they were compressed at the bottom of swamps or morasses. As a result, the usual aerobic process of rotting could not occur, and peat developed.

In the second phase, large amounts of heat and pressure were exerted on the peat-like material over hundreds of millions of years. The peat was covered by new oceans and land masses that developed over time. Geological movements of land masses, the creation of mountain ranges, and tectonic eruptions were required to generate today's coal resources. Through heat and pressure, water and other 'impurities' in the peat slowly volatized.

Through the coalification process biomass obtains the properties shown in Fig. 3.2. In the final stages, diamonds develop (Krüger 2007). Today's lignite

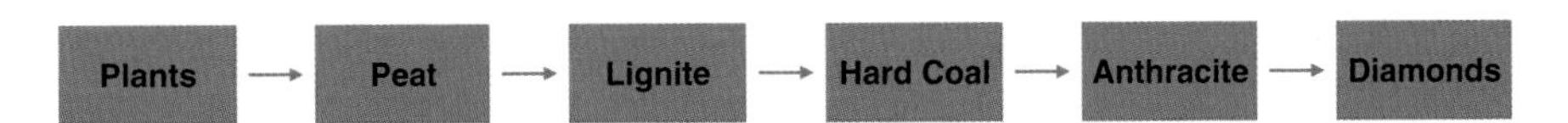

Fig. 3.2 Coalification process (Source: Author)

resources first developed in the Tertiary era about 65 to 2 million years ago, while diamonds that surface today are estimated to be between 1 and 3 billion years old.

Oil and gas differ from coal in their generation. Whereas coal developed mostly from terrestrial plants, most geologists support the biogenic theory in which oil and gas developed from small life forms and other ancient organic material. Here also compression and heat under oxygen-free conditions were required for the process to take place. This process also started over 300 million years ago.

3.3 Classification and Key Characteristics of Coal

This chapter classifies coal and summarizes its key characteristics. The purpose of this chapter is not to go into every detail of the scientific classification or chemical properties of coal but rather to summarize the knowledge that is required to better understand the coal market. Thus, this chapter examines the topic from an economical point of view to answer the scientific questions this book poses. I will therefore detail key characteristics of coal with respect to its use in the power industry. The reader may choose to selectively read over some of the passages below and refer to them later as required.

Ever since coal replaced wood-based biomass as the primary source of energy in the early 1800s, various classifications of coal have been developed. Today, coal is generally divided into low-rank coal and high-rank coal, the latter of which is often referred to as hard coal.

Figure 3.3 provides a good overview of the general classification of coal products. Here hard coal includes all bituminous coal and anthracite. Today, however, some subbituminous coal is also classified as hard coal (see here Indonesian subbituminous coal, which is part of internationally traded steam coal).

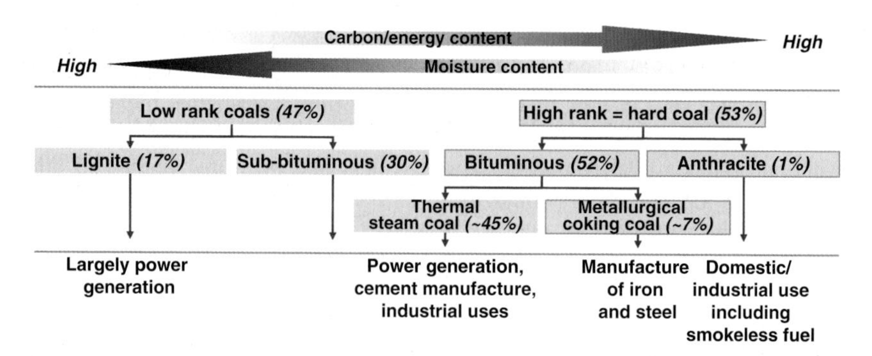

Fig. 3.3 Coal classification: general overview (Note: XX% = approximate percentage share of proven, probable, or indicated world coal reserves. Source: European Association for Coal and Lignite (EURACOAL); Author's analysis)

Coal Types and Peat					Total Water Content (%)	Energy Content af* (KJ/kg)	Energy Content (kcal/kg) NAR	Volatiles maf** (%)	Vitrinite Reflection in oil (%)
UN - ECE	USA (ASTM)	Germany (DIN)							
Peat	Peat	Torf			75	6,700	1,600		
Ortho-Lignite	Lignite	Weichbraunkohle			35	16,500	3,950		0.3
Meta-Lignite		Mattbraunkohle		HARTKOHLE	25	19,000	4,500		0.45
Subbitum. Coal	Sub-bituminous Coal	Glanzbraunkohle			10	25,000	6,000	45	0.65
Bituminous Coal	High volatility Bituminous Coal	Flammkohle	Steinkohle					40	0.75
		Gasflammkohle						35	1
	Medium Vol. Bitumin. Coal	Gaskohle				36,000	8,600	28	1.2
		Fettkohle			Kokskohle			19	1.6
	Low Vol. Bitumin. Coal	Esskohle						14	1.9
Anthracite	Semi-Anthracite	Magerkohle			3	36,000	8,600	10	2.2
	Anthracite	Anthrazit							

Fig. 3.4 International coal classification: detailed overview (Note: *af = ash-free; ** daf = dry ash-free. Source: European Association for Coal and Lignite (EURACOAL), Author's analysis

Figure 3.4 provides a more detailed overview of various US-, UN-, and German-based classification methods, including their official name. The classification of coal is generally derived from the key coal characteristics: (a) energy content or calorific value (in the coal industry indicated by the abbreviation CV); (b) total moisture content (indicated by the abbreviation TM); and (c) volatile content.

In this book, I refer to hard coal and steam coal. Steam coal is usually treated as a subsection of hard coal. Steam coal for our purposes includes anthracite (even if not used in the power industry) but excludes coking coal. Hard coal includes all coking coal. However, many industry specialists regard hard coal as excluding subbituminous coals. For the purpose of this study, subbituminous coal is included in the category of steam coal. In later chapters when I refer to reserves and resources I will only refer to hard coal since it is currently impossible to differentiate coking coal and noncoking coal or steam coal when discussing reserves and resources.

The following chapters summarize the key physical and chemical characteristics that are often referred to in steam coal supply contracts and communication between consumers, producers, and traders. The information below, though not complete, gives us a sense of coal's complexity as a commodity. Financial traders who enter the relatively new coal derivatives business and in some case the physical coal trade business often underestimate the complexity of the physical underlying asset, coal. In general, market participants with imperfect market knowledge tend to cause higher price volatility and underestimate risks. Lack of such knowledge can also incur high economic costs for an organization. Therefore, a basic knowledge of the key characteristics of coal is crucial (or at least highly recommended) to effectively participate in the coal market.

3.3.1 Calorific Value

The calorific value (or CV) is stated in kcal/kg or MJ/kg (equal to GJ/t). The lower heating value (LHV) – also known as net calorific value, net CV, or NCV – of a fuel is defined as the amount of heat released by combusting a specified quantity of product (initially at 25 C or another reference state) and returning the temperature of the combustion product to 150 C. The NCV assumes that the latent heat of the vaporization of the water in the fuel and the reaction products are not recovered. It is useful in comparing fuels when the condensation of the combustion products is impractical, or heat at a temperature below 150 C cannot be utilized. The NCV is generally used in the European coal trading business, which dominates the Atlantic coal trading market (for more detail, please refer to Section 4.2.2.4). Coal traded in the Atlantic market typically reaches a CV of 6,000 kcal/kg net as received.

By contrast, the gross CV (GCV) or higher heating value (HHV) includes the heat of condensation of water in the combustion products. The GCV is generally used in the Asian coal trading business, which dominates the Pacific coal trading market. Coal traded in the Atlantic market for Asian customers typically reaches GCVs of 5,500–6,700 kcal/kg gross air-dried. To convert gross into net or as received into *air-dried*, please refer to Table 3.2 on page 30.

Table 3.2 Converting coal characteristics: AR, adb, db, and daf

To obtain multiply	As received (AR)	Air dried basis (adb)	Dry basis (db)	Dry ash free (daf)
AR by		(100–IM%)/ (100–TM%)	100/(100–TM%)	
adb by	(100–TM%)/ (100–IM%)		100/(100 – IM%)	
db by	(100–TM%)/100	(100–IM%)/100		100/(100–A%)

where IM is the inherent moisture; TM is total moisture; A is ash.

To summarize, the NCV is the 'net energy' contained in coal that the power plant can use to generate electricity. The GCV is the 'total energy' that coal contains. However, part of this energy is required to vaporize the water in the coal when generating heat.

The following formulae, especially the simplified ones, are used in the coal industry to convert gross into NCV or NCV into GCV:

$$\mathrm{NCV(AR)} = \mathrm{GCV(AR)} - 50.6\,\mathrm{H} - 5.85\,\mathrm{TM} - 0.191\,\mathrm{O} \tag{3.3}$$

$$\text{Simplification 1}: \mathrm{NCV(AR)} = \mathrm{GCV(AR)} - 6(9\mathrm{H} + \mathrm{TM}) \tag{3.4}$$

$$\text{Simplification 2}: \mathrm{NCV(AR)} = \mathrm{GCV(AR)} - 260/300\ \mathrm{kcal/kg} \tag{3.5}$$

where H is the hydrogen content; TM is total moisture content; O is oxygen content.

Note: Simplification 2 assumes typical bituminous coal with 10% TM and 25% volatile matter

The most frequently used simplified formulae for converting steam coal from the Atlantic market standard net as received, NAR or NCV (AR), into the Pacific market standard gross air-dried, gad or GCV (adb), and vice versa are summarized below:

$$\mathrm{NCV(AR)} = \lfloor 1 - (\mathrm{TM} - \mathrm{IM}) \times \mathrm{GCV(ad)} \rfloor - 260\,\mathrm{kcal/kg} \tag{3.6}$$

$$\mathrm{NCV(AR)} = \mathrm{GCV(ad)} - 550/600\,\mathrm{kcal/kg} \tag{3.7}$$

where IM is the inherent moisture; TM is total moisture.

It should be noted that the CV is the key characteristic in coal trade. This seems obvious, but the CV is not as simple as it first appears. We must remember that the coal customer buys energy content per ton delivered to its power plant. Thus, higher CV will not only reduce the relative cost of transportation per ton of coal transported but also have an impact on the efficiency reached in the power plant when measured per ton of input product. However, an overly high CV may result in overly high temperatures in the boiler and therefore cause technical problems. Even some veterans in the coal trading industry often forget these very important facts about CV.

There is a general trend toward lower CVs when considering international coal supplies. Indonesian export volumes have increased significantly and often are of lower CV content. South African–exported coal products have also dropped in CV due to geological circumstances as well as due to wash plant capacity (in South Africa all exported coal is washed; for more details on coal washing, see Section 3.6.4). Coal from Russia has also dropped in CV due to geological reasons. In the long term, I expect that exported coal will continue to drop in CV. This will increase the relative cost of transportation and also pose challenges for old and new power plants. If everything else remains equal, the falling CV will result in higher prices per delivered ton of coal.

3.3.2 Ballast: Moisture and Ash Content

The key components in coal that do not carry calorific value are the moisture content and ash. Both together are referred to as ballast. The higher the ballast, the lower the carbon content and therefore the calorific value, and vice versa.

3.3.2.1 Moisture

All mined coal is wet. When coal is mined the product will include groundwater and other extraneous moisture, also referred to as 'adventitious moisture.' The moisture that is held within the coal itself is called 'inherent moisture' (IM). The total moisture content, both adventitious and inherent, is called just that: 'total moisture' (TM). The total moisture is all the water that the coal contains and, by definition, can only exist on an 'as received' (AR) basis since it cannot be air-dried (ad) or dried (db). By definition, inherent moisture can only be expressed on an air-dried (ad) basis, since any moisture that is removed through air drying is adventitious moisture. Coal with high inherent moisture content, such as Indonesian subbituminous coal, often soak up water after long periods of sunshine.

Coal that is crushed to smaller sizes tends to contain higher moisture than larger, uncrushed coal. This may be explained by the fact that extraneous water can flow off larger pieces of coal while it will be trapped more easily in fine coal (producing coal sludge in extreme cases).

Coal types with a very high moisture content, such as lignite, which can have total moisture content in excess of 50%, are often subjected to a drying process. Hard coal is rarely dried, as the inherent moisture cannot be dried out effectively and the extraneous moisture cannot be controlled during uncovered transportation, for instance on bulk vessels or trains.

Typical moisture content for seaborne traded steam coal with a calorific value of about 6,000 kcal/kg NAR varies from 8 to 15% AR (weight percentage). In general, one can summarize the impact of moisture on power plants as follows:

- Overly high moisture: Moisture needs to be removed, thus increasing the power plant's energy consumption. Moisture needs to be transported, increasing the relative cost per unit of energy. Moisture may also result in problems when handling frozen coal.
- Overly low moisture may cause problems with dust.

3.3.2.2 Ash

Ash is an inorganic matter (i.e., sand crystals). It is the residue that remains after coal is burnt. Thus, ash is noncombustible. It represents the bulk mineral matter after carbon, oxygen, sulfur, and moisture (including from clays) have been driven off during the combustion process. High ash contents in bituminous coals can be 'washed out' in a chemical and mechanical process called 'coal washing.' For instance, almost 100% of South African export coal (about 65 million tons per annum) is washed before shipping. Indonesian coal, on the other hand, tends to have very low ash content.

Typical ash content for seaborne traded steam coal with a calorific value of about 6,000 kcal/kg NAR varies from 6 to 15% AR (weight percentage). In contrast and by way of example, unwashed or untreated South African coal tends to have ash content in excess of 20% AR (for a discussion of coal washing, please see Section 3.6.4). Such high-ash coal would not be marketable internationally simply because the resulting low calorific value is not competitive with other coal sources.

The ash content can be expressed on an as received, air-dried, or dried basis. In general, one can summarize the impact of ash on power plants as follows:

- Overly high ash: Ash needs to be transported, thus increasing the relative cost per unit of energy. High ash content increases the capacity requirement of the flue gas cleaning equipment.
- Overly low ash content usually results in an increased share of unburned coal contained in the ash, thus reducing the ash quality. Ash is often sold to industrial users and unburned coal in ash makes the ash unmarketable.

3.3.2.3 Ash Fusion Temperatures (AFT)

When considering ash in coal it is also important to understand fusion temperatures. Ash is a noncombustible component. However, at very high temperatures (i.e., above 1,000 C) ash or the crystals held within it will melt. The temperature at which ash melts is called 'ash fusion temperature.' One differentiates between four different ash fusion temperatures summarized in Table 3.3.

Table 3.3 Four types of ash fusion temperatures

Name	... is the temperature at which
Initial deformation temperature (IT or T1)	The point of the ash cone begins to round
Softening temperature (ST or T2), also sometimes called the spherical temperature	The base of the ash cone is equal to its height
Hemispherical temperature (HT or T3)	The base of the ash cone is twice its height
Fluid temperature (FT or T4)	The ash cone has spread to a fused mass no more than 1.6 mm in height

The reason AFTs are important is that molten ash can stick to the inside of the boiler in the power plant and cause problems. Without going into too much detail, low AFT coal that has been burnt in boilers at temperatures above the fluid temperature T4 can result in large cleaning efforts and power plant outages. Especially older, less-efficient power plants that require high temperatures often have very restrictive AFT requirements of, for example, no less than 1,300 C T1 in reducing atmosphere.

The ash fusion temperature can be measured in a reducing atmosphere (red. atm.) or an oxidizing atmosphere (oxid. atm.). Generally, a temperature under reducing conditions should be equal to or lower than the corresponding temperature under oxidizing conditions. The difference in these temperatures generally increases with increase in iron content in the ash. Examples of ash fusion temperatures of selected US coal products are given in Table 3.4.

Measurement of AFTs is not 100% precise; therefore, tolerances of up to 40 K are usually accepted in international coal supply contracts when temperature differences arise between power plant requirements and the coal delivered.

The composition of the ash is determined in a separate ash analysis that many power plants require before burning a mine's coal for the first time. Differences in the ash composition of coal affect slagging and fouling behavior. Certain

Table 3.4 Ash fusion temperatures of selected US coal in °C

Fuel name	Initial deformation, reducing	Fluid, reducing	Initial deformation, oxidizing	Fluid, oxidizing
Eastern Kentucky	1,620	1,649	1,627	1,649
San Miguel Lignite	1,287	1,515	1,281	1,517
Pittsburgh #8	1,154	1,360	1,356	1,429
Upper Freeport	1,192	1,336	1,339	1,429
Illinois #6	1,060	1,253	1,241	1,366
Hanna Basin	1,196	1,281	1,246	1,344
Kentucky #11	1,076	1,268	1,268	1,341
Beulah Lignite	1,108	1,199	1,255	1,309
Roland	1,130	1,198	1,167	1,244
Black Thunder	1,161	1,210	1,179	1,243
Eagle Butte	1,217	1,245	1,196	1,231

Source: Brigham Young University, College of Engineering and Technology.

generalizations can be made about the influence of the ash composition on the ash fusion characteristics (IEA Clean Coal Center 2008):

- The nearer the ratio of Al_2O_3 and SiO_2 approaches that of alumina silicate, $Al_2O_3{\cdot}2SiO_2$ ($Al_2O_3 = 45.8\%$, $SiO_2 = 54.2\%$), the more refractory (infusible) the ash will be, thus increasing the AFT.
- CaO, MgO, and Fe_2O_3 act as mild fluxes, lowering the AFT, especially in the presence of excess SiO_2.
- FeO and Na_2O act as strong fluxes, lowering the AFT.
- A high sulfur content (from pyrite) lowers the initial deformation temperature and widens the range of fusion temperatures.

3.3.3 Volatile Matter

The volatile matter (also known as volatiles, volatile content, or VM) in coal comprises those components of coal, except for moisture, which are released at high temperature in the absence of air. VM is usually a mixture of short- and long-chain hydrocarbons, aromatic hydrocarbons, and some sulfur. The volatile content is a key indicator of how the coal will burn and what characteristic the flame will have. For instance, high-volatile coal tends to burn more quickly with a larger flame. Also, the coal self-ignites much more quickly. Low-volatile coal, on the other hand, tends to burn more slowly with a lower flame, but also at higher temperatures.

The volatile content can also be expressed on an as received, air-dried, or dried basis. Typical volatiles of steam coal range from 20 to 40% AR. For example, Indonesian coal tends to have around 40% AR (therefore, it also self-combusts faster) and South African coal tends to have around 25% AR. Russian and Colombian coal tend to have 30–35% volatiles AR. High-volatile coal with volatiles around or above 40% AR can slowly de-volatize over time when exposed to oxygen. Since the volatiles also contain hydrocarbons, this also means that the calorific value of high-volatile coal can decline over time. This is especially a risk with lower calorific material, such as subbituminous or younger coal. The ability to store low-CV/high-volatile coal is also significantly reduced and such coal needs to be burnt soon after it has been mined.

In general, one can summarize the impact of volatiles on power plants as follows:

- Overly high volatiles increase the risk of self-ignition and fires at the place of storage and inside the mills of the power plant.
- Overly low volatiles can cause problems with the stability of the flame at low loads and may also result in increased NO_x (nitrogen oxide) emissions.

3.3.4 Fixed Carbon Content

The carbon content of coal (called the ultimate carbon content) is responsible for the energy contained. The fixed carbon content of coal does not equal the ultimate carbon content. Fixed carbon is the carbon that is left after volatile materials are driven off. This differs from the ultimate carbon content of the coal because some carbon is lost in hydrocarbons along with the volatiles. Fixed carbon (also FC) is used as an estimate of the amount of coke that will be yielded from a sample of coal and is therefore especially relevant for anthracite products or coking coal. The fixed carbon content is only of marginal importance for the power industry as the calorific value, ballast, and volatile matter describe the coal well enough for power generation purposes. The ultimate carbon content, however, will determine the amount of CO_2 generated when burning the coal. Thus, the CV to C-content ratio is crucial when sourcing coal with the least CO_2 emission per MWh produced. Fixed carbon is determined by removing the mass of volatiles determined by the volatility test, above, from the original mass of the coal sample. Often, fixed carbon is determined by difference.

$$\mathrm{FC(db)} = 100 - \mathrm{VM(db)} - \mathrm{Ash(db)} \quad (3.8)$$

Please note that formula (3.8) works only on a dried basis and not on an as received or air-dried basis. In order to convert FC(db) one uses the formulae indicated in Table 3.2 on page 30.

3.3.5 Sulfur Content

The sulfur content of coal is the last of the key coal characteristics relevant to the steam coal market. Coal quality and sulfur content are usually inversely related to each other. However, since most power plant consumers often blend low-sulfur coal with high-sulfur coal, it results in a better market potential for higher sulfur material as well. Coal with sulfur content above 1% AR is classified as high-sulfur coal. Typically, high-sulfur material is discounted by about US $1 per 1% of sulfur above 1% AR, but discounts can vary significantly.

Less important for steam coal, but relevant for coking coal, is the fact that sulfur comes in organic and inorganic, usual pyritic (FeS_2), forms. In addition to pyrite it may also contain marcasite and sulfates, though the sulfate content is usually low unless the pyrite has been oxidized. The forms of organic sulfur are less well established and organic sulfur cannot be removed by physical means; existing chemical processes for removing organic sulfurs are usually very expensive.

In general, one can summarize the impact of sulfur on power plants as follows:

- Overly high sulfur content: The SO_2 formed may cause corrosion and environmental problems. High-sulfur coal therefore requires a larger desulfurization capacity when treating the flue gas. This results in more limestone required for the wet flue gas desulfurization equipment (WFGD equipment) and, thus, higher costs.
- Overly low sulfur content may reduce the performance of the electrostatic precipitator (ESP). Also, some German power plants are known to require a sulfur content of at least 0.5% AR on average in order to keep their desulfurization running efficiently, and to fulfill their long-term gypsum supply contracts. For reference, Russian steam coal usually comes at 0.2–0.4% AR sulfur content.

Figure 3.5 shows an extract from a typical analysis result of a shipment of South African coal loaded in December 2007 in Durban, South Africa. The analyses often display the results in all three basic forms: AR, ad, and db, as is also the case here.

RESULTS OBTAINED:

Determination		As Received Basis	Air Dry Basis	Dry Basis
Total Moisture, %	:	4.80		
Inherent Moisture, %	:		1.9	
Volatile Matter, %	:	18.7	19.3	19.7
Ash, %	:	15.6	16.1	16.4
Fixed Carbon, % (by diff.)	:	60.9	62.7	63.9
		100.00	100.0	100.0
Total Sulphur, S %	:	1.06	1.09	1.11
Gross Calorific Value, MJ/kg	:	26.83	27.65	28.19
Gross Calorific Value, kcal/kg	:	6409	6605	6734
Net Calorific Value, kcal/kg	:	6223		

Note: Analysis according to ISO analytical procedures unless otherwise stated

Fig. 3.5 Extract from an SGS steam coal analysis according to ISO

3.3.6 Size, Grindability, Nitrogen, Chlorine, and Fluorine

The size of the coal in millimeters is another important factor for hard coal. Steam coal for power generation is usually offered in the size 0–50 mm with no more than 5% above 50 mm. Since modern power plants work with pulverization, in theory one could even supply 0–6 mm coal; however, here the risk of dust and environmental problems is very high. The grain size affects the transportation on belts and the choice of mills in the power plant. Sized coal (separated through sieves) is traded at a significant price premium of around 10–30% compared to similar 'fine' coal. Such sized coal is used for domestic purposes and industrial uses where some older ovens can work only with sized material. Run-of-mine coal (ROM or coal straight from mining without crushing) usually has the size 0–300 mm. However, surface mining can result in larger junks of up to 500 mm in size.

The grindability is expressed through the hard grove index (HGI). The HGI determines how hard the coal is. High HGI scores indicate soft coal while low HGI scores indicate hard coal. Harder coal (i.e., HGI below 40) is more difficult to grind. As a result, larger coal particles need to be supplied to the boilers. This, in turn, may result in an increased amount of unburned coal in the fly ash.

Nitrogen content is responsible for the nitrogen oxide emissions of power plants. Nitrogen oxide is especially carefully monitored in the United Kingdom and the United States. The relationship between coal nitrogen and emissions of nitrogen oxides is not clearly understood to date; this is an area in which further research is needed (Davidson 1994).

Chlorine and fluorine contents affect the performance of the wet flue gas desulfurization and the gypsum quality. The risk of corrosion increases with higher chlorine and fluorine values. Chlorine is one of the most troublesome components of coal in combustion applications, causing slagging, fouling, and corrosion. There is substantial evidence that fouling and corrosion increase as the chlorine content in coal increases above 0.25–0.5% AR (IEA Clean Coal Center 2008).

3.4 Regional Analysis of World Coal Production, Reserves, and Resources

The world's coal resources are widely distributed. One of the major advantages of coal compared to other fossil energy sources is that resources are available not only in developing countries but also in Europe, Australia, the United States, and other countries that are generally considered part of the western and more stable world. This is exemplified by Australia's status as the world's largest coal exporter (counting steam coal and coking coal together), accounting for about one-third of world seaborne hard coal exports.

For the purpose of global overview, in this chapter I have selected the data published by the German Federal Institute for Geosciences and Natural Resources (*Bundesanstalt für Geowissenschaften und Rohstoffe*) in 2006 (BGR 2006). I have selected this source because of its reliability and because it also provides

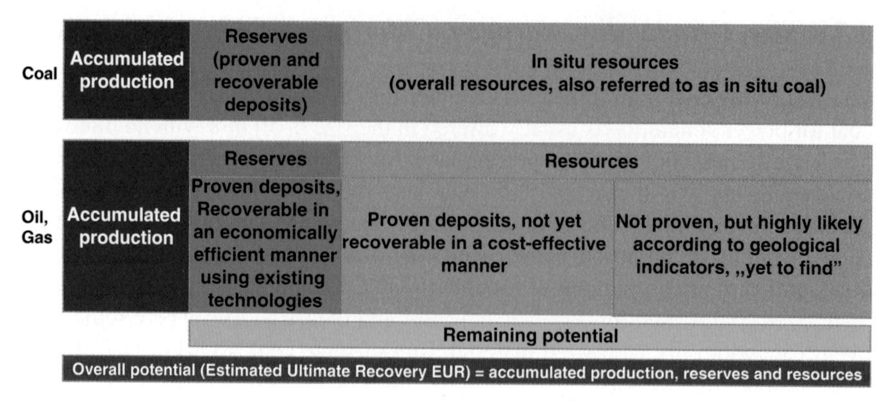

Fig. 3.6 Definition of coal, oil, and gas reserves and resources (Note: The definition may vary depending on source used. Source: IEA-Manual 2006)

comparable data for oil, gas, and uranium. In this chapter, I will focus solely on the energy resource coal and the world coal market. I shall use the internationally accepted definitions of reserve and resource as detailed in Fig. 3.6.

3.4.1 World Resources, Reserves, and Production Analysis

The world has very large coal reserves and resources. Figure 3.7 summarizes the global distribution of hard coal. While the total production in 2006 was 5.4 billion

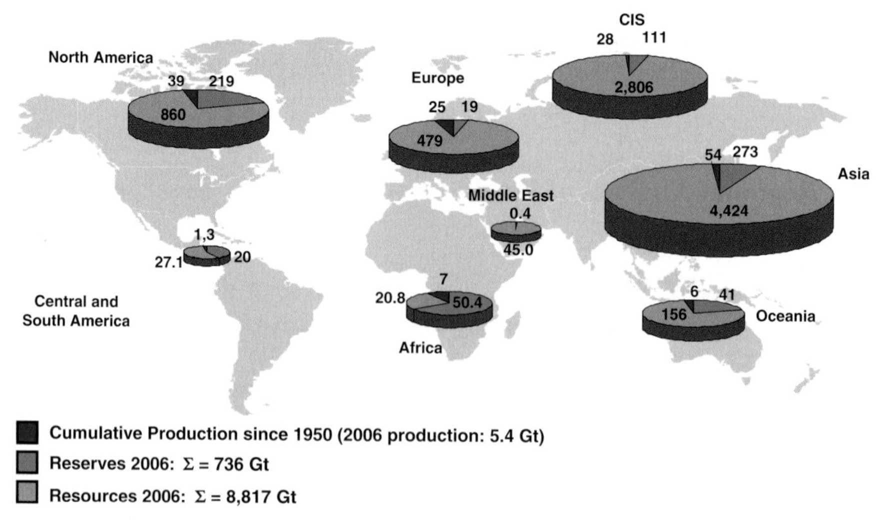

Fig. 3.7 Global hard coal production, reserves, and resources (Source: BGR 2006; Author's research and analysis)

tons, of which about 800 million tons were exported via sea, total hard coal reserves are estimated at 736 billion tons (lignite: 283 billion tons), resulting in a theoretical reserve/production ratio for hard coal of 137 years. The total remaining potential, combining reserves and resources, is about 9,550 billion tons of hard coal (lignite: 3,356 billion tons), resulting in a remaining theoretical potential/production ratio for hard coal of about 1,780 years (BGR 2006). However, when interpreting these numbers it must be remembered that coal reserves and resources do not differentiate between raw coal and sellable coal. In many countries such as Australia and South Africa where the coal is washed, only 65–80% of raw coal translates into sellable coal. Kjaerstad and Johnsson (2008) argue that the reserve/production ratio is therefore meaningless. While I agree that this factor makes the figures less reliable, they nonetheless demonstrate that coal is available in abundant quantities, which is the point I wish to discuss here.

Current production has only scratched the surface of the remaining potential of coal. The largest potential (reserves plus resources) in tons of hard coal exists in China and Russia. The reserves of the United States are the largest in the world, which is most likely a function of the investments in and professional surveys done on US coal reserves (Fig. 3.8).

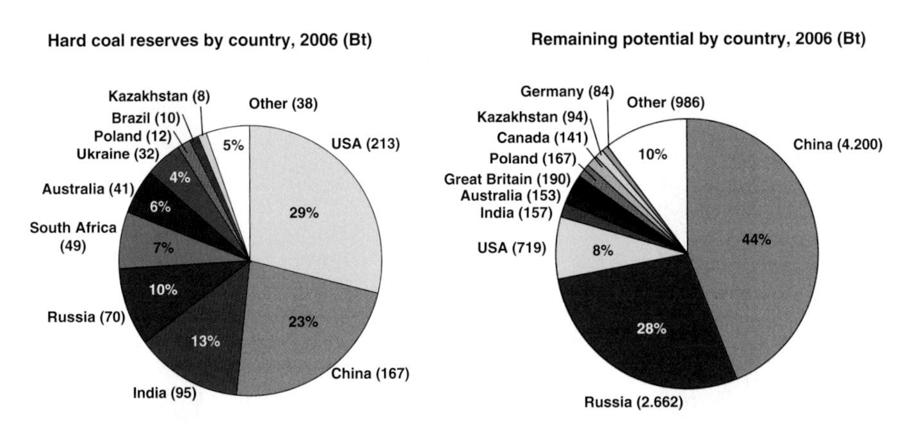

Fig. 3.8 Global reserves and resources by country, 2006 (Note: resources + reserves = remaining potential. Source: BGR 2006, Author's analysis)

Reserves are well distributed across the globe, with countries in Asia, Europe, North America, and South America in the top 10. This fact is good news for the international coal market, in particular for steam coal consumers, since diversified sources translate to greater reliability of supply. Table 3.5 presents the 10 countries with the largest reserves. Indonesia is probably the most interesting case. According to the BGR, if it continues to produce at the current rate it will have exhausted its reserves in only 14 years. This, of course, will not happen. It indicates the difficulty with such statistics where the availability of data is limited. In fact, Indonesia is expected to have well in excess of 50 years' reserves, but it is true that

Table 3.5 Hard coal reserves and resources of top 10 countries by reserves

No.	Country	Prod. 2006 (million tons)	Reserves (million tons)	Percent	Resources (million tons)	Percent	Remaining potential (million tons)	R/P (years)	RP/P (years)
1	United States	999	213,316	29	718,939	8	932,255	214	933
2	China	2,381	167,000	23	4,200,000	48	4,367,000	70	1,834
3	India	399	95,399	13	156,960	2	252,359	239	633
4	Russia	233	69,946	10	2,662,155	30	2,732,101	301	11,741
5	South Africa	247	48,751	7	n/a	0	48,751	197	197
6	Australia	305	40,800	6	152,900	2	193,700	134	635
7	Ukraine	80	32,039	4	49,006	1	81,045	402	1,017
8	Poland	96	12,459	2	167,000	2	179,459	130	1,879
9	Brazil	6	10,113	1	15,319	0	25,432	1,605	4,037
10	Kazakhstan	93	8,000	1	94,000	1	102,000	86	1,097
	–								
11	Colombia	64	6,611	1	4,600	0	11,211	103	175
	–								
15	Indonesia	200	2,890	0	23,949	0	26,839	14	134
	–								
43	Germany	24	99	0	84,474	1	84,573	4	3,599
	–								
	Total	**5,357**	**736,112**		**8,817,303**		**9,553,415**	**137**	**1,783**

Note: R/P = reserve/production; RP/P = remaining potential/reserves Source: BGR (2006, p. 66).

Indonesia's reserves are relatively modest considering its status as the largest steam coal exporter in the world. From Table 3.5 we can determine that the resource/reserve ratio for hard coal is 12:1. This ratio was 5:1 two years ago according to Ritschel and Schiffer (2007, p. 10). Thus, even though resources are reassigned to reserves, relatively speaking, known resources have increased at extraordinary pace and are expected to continue to do so.

In general, the estimates of coal reserves and resources are subject to continuous adjustment, similar to the estimates concerning oil and gas. However, while figures for oil and gas are systematically updated, the same has not been true for coal deposits. The international community has historically spent much less time and capacity updating data on coal deposits than those on oil and gas. Also, relatively little effort has gone into researching new and long-term coal reserves. We can therefore expect that figures for coal reserves and resources will increase significantly once more time, effort, and money are spent on research. At the same time, as coal prices rise, reserves will be reassigned to resources, also increasing the reserve/production ratio R/P.

Hotelling (1931) has worked out the special case of depletable resources in his well-known paper 'The Economics of Exhaustible Resources.' If Hotelling is right, then all minerals in situ must inevitably and progressively become scarcer and more valuable. However, Adelman and Watkins (2008) have concluded in their paper 'Reserve Prices and Mineral Theory' that "there is an endless tug-of-war between diminishing returns and increasing knowledge" when it comes to reserves and resources. This supports scientifically the experience in the coal industry where new coal resources are constantly being discovered while production depletes them. This, according to Adelman and Watkins, explains why for many decades it appeared that fossil fuels would last forever, since their 'life expectancy' did not decline. However, even if fossil fuels sometimes appear to be endless, in fact they are not.

The following sections will provide further detail about the main exporters production of seaborne steam coal: Australia, Indonesia, South Africa, Russia, China, and Colombia. Table 3.6 summarizes the importance of coal export countries in the global arena.

Table 3.6 World share of coal exports, 2006

Country	Share of global steam and coking coal exports (%)	Share of global steam coal exports (%)	Share of global coking coal exports (%)
Australia	32	20	67
Indonesia	20	26	0
Russia	10	12	5
South Africa	9	12	1
China	9	11	2
Colombia	8	10	0
Other	12	9	25

Source: Kopal (2007, p. 16).

3.4.2 Australia

Overview: Australia has long been one of the world's key suppliers of minerals and energy raw materials. In 2006, 20% of the world's exported steam coal was sourced from Australia (Kopal 2007). At current production levels, Australia can continue for 137 years utilizing its reserves and 635 years if it were to utilize its remaining potential (BGR 2006). From a global perspective, Australia's role in the supply of coking coal is even greater, as two-thirds of the world's seaborne coking coal are sourced from Australia.

3.4.2.1 Australia's Production and Exports

Australia's coal reserves are located in the east of the country in the states of Queensland (2006 total production: 177 million tons) and New South Wales (2006 total production: 128 million tons, Ritschel and Schiffer 2007). Of the total production of 314 million tons, about 237 million tons or 78% were exported. Ritschel and Schiffer (2007) estimate production slightly higher than BGR (2006). Total export capacity of all export-oriented mines is estimated by Kopal (2007) at 297 million tons in total. Thus, Australia reached an export capacity utilization of 79%. For steam coal capacity alone, however, Kopal (2007) estimates this figure as reaching 82%. Therefore, export capacity utilization for Australia in 2006 was below the average global steam coal export capacity utilization of 88%.

Of the total production, about three-quarters are produced in opencast mines and one-quarter underground. There are about 100 mines in Australia. The coal is mined down to about 70 m in the opencast mines. Underground mines can reach depths of about 200 m. While Australia's mining operations are among the best in the world, productivity seems to have decreased, considering that in 2006 almost 35,000 people worked in the coal mining industry, up from 10,000 two years before (Ritschel and Schiffer 2007).

Australia's coal production is almost entirely privately owned. The Australian Coal Report estimates that the four biggest Australian producers – BHP, Rio Tinto, Xstrata, and Anglo – accounted for 53% of Australia's output and for 55% of the country's exports in 2006. The Australian coal industry is likely to continue to be consolidated. In addition, Chinese and other new Asian entries have started purchasing smaller mine assets and are expected to continue to do so (Table 3.7).

Australia is expected to develop further steam coal export mining capacity of 172 million tons by 2011, according to Kopal (2007). Australia is therefore estimated to account for almost 60% of all planned and agreed world steam coal export mining expansions (totaling 292 million tons) by 2011. However, the key constraint in Australia is the expansion of infrastructure capacity, namely rail and port infrastructure.

Table 3.7 Australia's most important hard coal producers

Company	No. of mines	Output 2006 (million tons)	Exports 2006 (million ton)
BHP-Billiton Ltd.	16	45	40
Rio Tino Ltd.	8	37	31
Xstrata PLC	21	54	40
Anglo Coal Australia Pty Ltd.	7	31	20
Total	**52**	**167**	**131**
Percentage of total		53	55
Australia 2006		314	237

Source: Australian Coal Report, output is pro rata.

3.4.2.2 Australia's Coal Quality

Australia's steam coal exports are known for their homogeneity and high quality. Since Australian coal is washed in a similar way to South African coal, the high ash content of the raw coal is reduced sufficiently to supply a reliable 6,000 kcal/kg NAR quality or even better. Water is scarce in Australian mining regions and the government has put a price on water. This may increase the cost of coal washing and as a result could lead to lower coal qualities in the future. Today, Australian coal has medium volatile matter (25–30% AR), average ash content (8–15% AR), and tends to be rather dry (7–9% AR). The sulfur content is below 1% AR and can go as low as 0.3% AR. The coal is also relatively soft (higher HGI). Overall, Australia's steam coal quality is well suited for the export market. Japan, the world's largest coal importer and also Australia's most important customer, is considered to be 'spoiled' by Australia's good coal qualities. International coal consumers, especially in Japan, have tended to prefer Australian coal over, for example, Indonesian and Chinese coal, and are therefore often willing to pay a premium, especially in Japan.

3.4.2.3 Australia's Infrastructure

Australia has been particularly hampered by a lack of infrastructure development. Both the rail lines from the mines to the ports and port capacity need to be expanded. During 2006 and 2007, at times over 60 vessels have queued in Newcastle (New South Wales), not only taking up sea transport capacity but also exposing the insufficient infrastructural capacity of Australian export ports.

Australia exports its coal resources through seven ports; in order of importance these are as follows: Newcastle, Dalrymple Bay, Gladstone, Hay Point, Abbot Point, Port Kembla, and Brisbane. All ports have expansion plans. Australia's port capacity in 2006 was about 268 million tons (signifying 88% capacity utilization in relation to 237 million tons of coal exports). Short-term expansion plans will increase the capacity to 325 million tons, and in the medium term the port capacity is expected to reach a capacity of over 450 million tons (VDKI 2006).

3.4.2.4 Australian Costs

Australia is one of the world's most efficient and therefore also cost-effective coal producers. The distance from the mines to the ports ranges from about 100–400 km, and geologically the coal is situated advantageously with undisturbed deposits. I estimate that Australia's 2006 FOB marginal cost for steam coal ranges from 21 to US $44/ton (see author's FOB marginal cost curve analysis, Section 5.4). This figure varies from US $21/ton for Queensland's most efficient surface mines to US $43/44/ton at New South Wales' smaller and less-efficient underground mines. Costs are likely to increase further for a number of reasons, including the following (VDKI 2006; Ritschel and Schiffer 2007):

- Coal from newer deposits will be railed longer distances.
- Overburden to coal ratios is expected to increase and drive up pure production costs.
- Machinery costs will go up because equipment suppliers lack capacity to fulfill global mining requirements (this also affects iron ore and other mineral production in addition to coal).
- Fuel costs will rise, causing mining costs to rise with them.
- Australian royalties are linked to the sales price; with increased international coal prices as in 2007 and 2008, the royalty cost will increase.
- Ash content of the coal will increase, reducing washing yields and exported qualities.

The above list of cost drivers is valid for almost every producing country in the world. For a discussion of the global steam coal supply curve, as well as marginal costs, please refer to Chapter 5.

3.4.3 Indonesia

Overview: In recent years, Indonesia has developed into the most important exporter of steam coal, overtaking Australia for the first time in 2005. In 2006, more than one-quarter of all globally traded steam coal was sourced from Indonesia. Indonesia's export volumes have grown at unprecedented rates. For example, in 2001 Indonesia exported only 67 million tons of steam coal. By 2006 this had increased to 171 million tons, representing a compound annual growth rate (CAGR) of 21% compared to a global growth rate of 8% for seaborne steam coal traded during the same time period (VDKI 2006). Indonesia is relatively new to the coal mining business, with modern coal mining only starting in the 1980s; according to the Indonesian Mining Association (Indonesian Mining Association 2006), in 1993 Indonesia's total coal production comprised only 27 million tons. Government policy up to now has prevented international standard consolidation in Indonesia's coal industry. Through a relatively complicated mining rights system the government distributed state-owned coal reserves in three tranches (referred to as Coal Contracts of Work, CCOW) in the early 1980s (Ritschel and Schiffer 2007). In 2008/2009 the government is expected to redraft the mining law. It is unclear what impact this will have on Indonesian exports.

The CCOW contractors undertake to prospect for and explore the coal deposits located in their concession area, possibly to engage in mining development and, in return, are granted exclusive rights for a term of 30 years subject to a royalty of 13.5% of revenue. The contractors are also obliged to offer Indonesian investors at least 51% of the mining stock after a 10-year operating period. For instance, Rio Tinto/BP and BHP-Billiton were affected in 2001 and had to reduce their shareholdings. Today, Indonesian companies control production. However, through joint ventures and off-take agreements, large international conglomerates and trading houses have access to the majority of Indonesian coal output. Today, in addition to these larger production operations hundreds of smaller miners with annual output of about 0.1–1 million tons per annum have developed and take advantage of mining rights granted by local governments. Local governments can grant exploration and exploitation licenses for areas of up to 5,000 ha without consulting the central government in Jakarta.

With about 250 million inhabitants, Indonesia is the world's fourth most populous country after China, India, and the United States. Indonesia's electricity demand is still rising rapidly. Historically, Indonesia – until May 2008 a member of OPEC – has relied largely on oil and gas for its electricity production. In the near future, Indonesia will switch to coal and increase its own coal consumption almost tenfold within the next 20 years. The Indonesian Coal Mining Association estimates that in 2006/2007 about 37% of Indonesia's 26-GW installed power generation was fueled by coal (Argus Coal Daily, 2 January 2008). PLN, the state-owned power generator, which controls more than 80% of Indonesia's power, plans to increase coal-fired power generation to over 60% after 2010. Considering that Indonesia's electrification ratio is only about 54%, this seems ambitious but possible.

3.4.3.1 Indonesia's Production and Exports

About two-thirds of Indonesia's coal reserves are located in East and South Kalimantan. The remainder is located in Sumatra and on other islands. While BGR (2006) estimated the reserves and resources at 3 and 24 billion tons, respectively (see also Table 3.5 on page 40), the newest figures disclosed in interviews with the Indonesian Mining Association totaled 7 and 58 billion tons in reserves and resources, respectively. Indeed, even these numbers are considered very conservative. Virtually all of Indonesia's coal production is opencast.

The top six coal producers control two-thirds of Indonesian exports (Table 3.8). However, foreign conglomerates have only limited control. Mining laws are often erratic; in early 2007, for instance, the government decided on a US $1.5/ton export tax to be implemented within 2 months. This tax was abolished the same year.

Table 3.8 Indonesia's most important steam coal producers

Company	Output 2006 (million tons)	Exports 2006 (million tons)	Notes
PT Adaro	34	34	Local
PT Kaltim Prima	34	25	Bumi Resources (local), marketing by Glencore, former Rio Tinto/BP principal owners: Bakrie and Tata Power in India
PT Kideco	18	19	46% Indika Energy (local), 49% Korean Samtan
PT Arutmin	16	16	Principal owners: Bakrie and Tata Power in India
PT Berau Coal	11	11	Local
PT Indominco Mandiri	9	11	Local
Total	**122**	**115**	
Percentage of total	59	67	
Indonesia 2006	205	171	

Source: Ritschel and Schiffer (2007); Author's market interviews.

Kopal (2007) estimates that Indonesian export producers worked at 96% capacity in 2006. This is supported by the price increases in 2007 and 2008. However, Indonesia is a difficult country to judge. Probably 25–35 million tons of output is not officially recorded. The total number of mining companies surpasses 500. Little is known about the large number of second- and third-tier producers with annual output of 0.1–2 million tons.

Total production is expected to increase from 205 million tons in 2006 to over 300 million tons in 2010 and subsequent years. The Indonesian Mining Association estimates that Indonesia's own consumption will increase from over 30 million tons in 2006 to approximately 100 million tons in 2010, 135 million tons in 2015, 185 million tons in 2020, and 220 million tons in 2025 (Indonesian Mining Association 2006). Indonesia will therefore have trouble maintaining its high export volumes. This difficulty coupled with already high capacity utilization will adversely affect the global seaborne steam coal market and possibly assert upward prices pressure, since the market has been fueled by Indonesia's export growth.

The five main countries that Indonesia supplies with coal are, in order of volume, Japan, South Korea, Taiwan, India, and Hong Kong. According to Ritschel and Schiffer (2007), these five destination countries accounted for 63% of Indonesia's exports in 2006.

3.4.3.2 Indonesia's Coal Quality

Indonesia's coal reserves are generally of lower quality than those of other exporting countries. Ash (1–9% AR) and sulfur (0.1–1% AR) contents are low, while total moisture contents are usually high (15–25% AR). The majority of the exported coal is bituminous; however, for local consumption much of the coal is subbituminous. The Southeast Asian market (i.e., the Philippines, Thailand, and India) is able to offtake some of Indonesian's subbituminous coal qualities with total moisture running as high as 35% AR, while Japan, China, South Korea, Taiwan, and Hong Kong generally prefer to buy coal of a slightly higher calorific value. Where freight rates allow, Indonesian coal is also transported to Europe. However, in the second half of 2007, for instance, exports to Europe virtually stopped due to strong Asian demand and high freight costs.

3.4.3.3 Indonesia's Infrastructure

Of the world's steam coal-exporting countries, Indonesia is probably the country with the fewest worries about infrastructure. This, however, by no means infers a perfect inland transportation and transshipment system. By nature and due to Indonesia being a country of many islands, Indonesia's coal mines are all located very close to the sea or rivers. Trucks are used to transport the coal to river or sea ports. There are about 15 coal-loading ports and about 20 offshore loading facilities (the so-called anchorage points where coal is transshipped via floating cranes or self-loading vessels from barges onto sea-going vessels). There are countless barge-loading facilities located along rivers or the seashore. Kalimantan has many rivers and almost every mine, even the very small ones, has its own barge port facility. Barge loading capacities range from 3,000 to 12,000 tons. Some exports to nearby countries are handled purely by barge.

Indonesia does not have official plans for infrastructure developments, but private companies have started to plan railway lines and new road projects. These projects are aimed at accessing Indonesia's very interesting high quality and even some coking coal reserves in Central Kalimantan, Sumatra, or other areas. Overall, due to the variety of loading facilities and due to the fact that most Asian coal-importing countries are only able to accept smaller-sized vessels, Indonesian coal production and export are not likely to be hampered by its infrastructure, as simple as it is.

3.4.3.4 Indonesia's Costs

Indonesia is probably the lowest-cost coal producer on a per-ton basis. The main reason for this – its inland logistics – has already been discussed above. Labor costs are also still very low with an average mine worker earning US $50–100/ month. There are no productivity data available, but the many smaller mines work much less efficiently on a per-worker basis than in the rest of the world. However, one can assume that the big six producers can compete with international productivity levels. The larger mines are managed by large Australian and other foreign mining contractors.

I estimate marginal 2006 FOB costs to be between US $20 and 35 per ton of coal. The big six are expected to reach US $20 in East Kalimantan, while very small and less-efficient mines in South Kalimantan and Sumatra will require US $35 per ton FOB mother vessel to marginally break even. Costs are expected to continue to increase. Since virtually all operation is carried out on a truck-and-shovel basis, fuel costs are a major issue, often accounting for more than 30% of total FOB costs. This figure may seem very high, but much of the electricity used during loading and mining is fuel-based and provided by diesel generators. Fuel costs used to be subsidized by the government, but in 2006 and 2007 subsidies were reduced, resulting in a 200–250% fuel price increase.

Specific one-time investment to develop new mines is relatively low at US $5–25/ton of annual output. Since this figure is far below the global average of US $50/ton of annual output (Kopal 2007), Indonesia is expected to cope with increased local demand while keeping its export levels at least constant, if not expanding them to about 200 million tons per annum.

3.4.4 South Africa

Overview: With a 12% share of world steam coal exports, South Africa is, alongside Russia, the third most important exporter of steam coal. However, for the Atlantic market (including Europe), South Africa – with a 27% share – is the most important exporter (Ritschel and Schiffer 2007, p. 25). South Africa has a special position in the international steam coal business for the following reasons: (1) it runs the largest and most efficient coal export port, Richards Bay Coal Terminal; (2) it is a price trend setter for Europe; and (3) the FOB South African price is quoted daily in the form of the API4 Index (see Abbreviations and Definitions), ensuring better market transparency.

For South Africa itself, the mining industry (and to a large extent coal) plays an above-average role for the health of the economy (Dlamini 2007):

- 16% of GDP (directly and indirectly);
- 50% of merchandise exports (primary and beneficial mineral exports);
- 20% of fixed investments (directly and indirectly);
- 35% of South African JSE's market capitalization;
- 25% of formal sector employment (directly and indirectly);
- 50% of rail and port volume;
- coal accounts for 90% of electricity generation and 15% of electricity demand; and
- coal accounts for 30% of liquid fuels from coal by Sasol, the world's market leader in CtL.

In a drive for a fairer and more efficient use of South Africa's natural resources, and as part of its Black Economic Empowerment (BEE) program, the South African government agreed to a new mining law in 2002. This law ensured that all of South Africa's natural resources were transferred from previous land ownership to state ownership. Thus, today, ownership of lands and mining rights can be separate. All previously operating mining companies had to reapply for their mining rights. New regulation also requires that older mining companies and new mining companies be partly black-owned. This has caused some problems due to an educational and financial gap between the black and white populations, but this gap is narrowing.

South Africa's reserves and resources are ample. With 49 billion tons of reserves (Table 3.5) and 115 billion tons of resources (BGR 2006) the country has almost 200 years of reserves and, additionally, over 600 years of resources at current production levels. The main reserves (about 20 billion tons) are located in the Highveld and Witbank areas just outside of Johannesburg.

3.4.4.1 South Africa's Production and Exports

BGR (2006) and VDKI (2006) estimate South Africa's total production at 247 million tons in 2006. The latter determined total exports of hard coal at 69 million tons (68 million tons of steam coal and 1 million ton of coking coal). In 2006, South Africa consumed about 176 million tons domestically: 109 million tons for power generation, 44 million tons for synthetic fuels, 18 million tons for industry and household use, and 5 million tons for the metallurgical industry. Of the 69 million tons exported, 52 million tons, or 76%, were shipped to Europe (Ritschel and Schiffer 2007, p. 67). India started to import more South African coal in 2007 and 2008.

South Africa's coal production is highly concentrated, with the top five producers controlling 86% of exports (see Table 3.9). However, due to new mining regulations and increasing involvement of BEE companies, the consolidation process is stagnating. It is, though, expected that BEE-controlled Exxaro will form alliances and take over other smaller BEE players, increasing its export share to 9–12 million tons within the next 3–5 years.

Table 3.9 South Africa's most important steam coal producers

Company	Output 2006 (million tons)	Exports 2006 (million tons)	Notes
Anglo Coal	59	19	
BHP-Billiton Plc.	52	21	
Sasol	47	4	Uses 43+ million tons for CtL
Exxaro	24	2	BEE, Former Kumba and Eyesizwe
Xstrata Plc.	21	13	35% Glencore-owned
Total	**203**	**59**	
Percentage of total	82	86	
South Africa 2006	247	69	

Source: Ritschel and Schiffer (2007), Author's market interviews.

Owing to infrastructural constraints (see below), the larger producers had slowed their investments. Kopal (2007) estimates export mining capacity in 2006 at 76 million tons, resulting in a tight 91% export production capacity utilization. Based on market interviews, it seems that Kopal's figure could be optimistic. A larger mining expansion push is expected in the near future up to 2010. Eskom, the national power company, requires larger coal input (see power crisis of South Africa with widespread blackouts in January 2008), and Richards Bay Coal Terminal is being expanded from 72 to 91 million ton capacity.

The outlook for South Africa is positive, as demand from the Pacific market, namely from India and even South Korea, is going to increase as the Pacific supply market tightens. In 2007 and 2008, South Africa was unable to participate in the worldwide coal export growth, but after Richards Bay Coal Terminal has finished expansion and after railroads improve in efficiency, South Africa will try to retain its strong position among the world's exporters.

3.4.4.2 South Africa's Coal Quality

South Africa produces a high-ash, low-moisture coal. All exported coal volumes are mechanically and chemically washed, reducing the ash content from around 25% AR to 10–15% AR. The washing yield (typically in the 60–80% range) determines not only the washing capacity required but also the marginal cost, since all costs up to washing have to be divided by the washing yield in order to determine the true costs of exported coal, as discard material (left over from washing) has little or no value. South African coal otherwise has a very good reputation on the market, because washing guarantees a very homogenous output product. Quality control at Richards Bay Coal Terminal is very tight; in fact, the two major coal qualities exported from RBCT are referred to as RB1 and RB2 quality (the only difference is in volatile matter, RB1 = min 22% AR volatiles and RB2 = min 25% AR volatiles). Thus, Richards Bay coal does not differentiate between coal from different mines.

3.4.4.3 South Africa's Infrastructure

As already mentioned above, South Africa operates the world's largest and most efficient coal export terminal – Richards Bay Coal Terminal – with a current capacity of 72 million tons (12% of the world's 2006 steam coal exports). RBCT is privately owned and was set up in 1976 with 12 million ton capacity. The expansion (known as Phase V) from 72 to 91 million ton capacity has already been agreed and is expected to be available from the second half of 2009, though many industry experts believe this timeframe to be unrealistic. Richards Bay Coal Terminal is currently 91% controlled by five shareholders: BHP Billiton (33.27%), Anglo (26.91%), Xstrata/Glencore (20.48%), Total (5.57%), and Sasol (4.90%) (RBCT 2008).

The government-owned railroad system in South Africa has been the key capacity constraint for higher exports. COALlink, which operates the 600-km-long railway line from the key mining areas to Richards Bay Coal Terminal, has a capacity of 72 million tons, not enough to ever reach the full port capacity. Trains running on this line can load more than 16,000 tons of coal in one load, for instance with 250 wagons of 64 tons each. Such trains are several kilometers long. The inefficiencies of a state-run rail operation also affect South Africa. While there is some talk about privatization, it is not clear when this will happen. Service philosophy and accountability are key problems at COALlink. One example that illustrates these problems was a presentation by a railway representative at a coal conference in Cape Town in January 2008, in which it was suggested that the railway should participate in the coal price increases at that time. He did not mention, however, that the railway is not willing to take any risks when coal prices are low. The deadlock between producers demanding more rail capacity and the railway demanding 20-year or longer guaranteed usage agreements will continue and may restrict South Africa's coal export expansion.

3.4.4.4 South Africa's Costs

South Africa remains a very cost-efficient coal producer when viewed in the global context. Despite increases in labor and material costs, based on detailed market analysis and primary research, my estimates of the country's 2006 FOB marginal cost vary from US $27 to US $48/ton. This figure is in line with international estimates such as those published by Ritschel and Schiffer (2007) and IEA Energy Outlook (2007). The new, smaller mining companies are, of course, producing at the higher rate, while the majority of South Africa's coal is produced at the lower cost rate.

All of South Africa's production is opencast. Western mining companies have established very efficient production operations. Inland transportation remains relatively inexpensive, and with Richards Bay Coal Terminal, the marginal transshipment cost for the mining companies is very low. The future cost of new BEE laws is difficult to estimate. In my discussions with various mining executives, it remained an unidentified 'cost problem' that in the long term should not be relevant for South Africa's global competitiveness in the coal market.

3.4.5 Russia

Overview: Russia, like South Africa, accounts for 12% of global steam coal exports (see Table 3.6). With a 23% share, it is the third most important coal source for Europe after South Africa and Colombia (Ritschel and Schiffer 2007, p. 25). Russian coal has a long history, though not in the international arena. Russia's own industry consumes the majority of the country's production output. In the early 1990s following the collapse of the Soviet regime, small quantities of Russian coal started to appear on the international seaborne steam coal market. Russia's export growth really began only after 1997, but was not as spectacular as Indonesia's. The Russian coal industry was very quick to privatize in the 1990s, resulting in an immediate cut in large amounts of unprofitable production. In fact, Russian coal production was cut by over 35% from 401 million tons in 1989 to 272 million tons in 1994 (see Global Insight-World 2007).

Russia's reserves and resources are among the largest in the world. The country accounts for 30% of the known coal resources (surpassed only by China with 48%). Its 70 billion tons of reserves and 2,662 billion tons of resources give Russia the largest remaining production potential at almost 12,000 years (BGR 2006). Russia is one of the most important steam coal exporters. Europe, and increasingly Asia, relies on Russia's logistical proximity. Russia is especially important for the economics of the coal market since it is the key marginal cost supplier. Thus, long-term, global FOB prices will be close to Russia's FOB marginal cost. (For a detailed discussion about the economics, please see Chapter 6.)

Russia already consumes 171 million tons, or 72% of its 239 million annual tons of steam coal domestically (VDKI 2006; Ritschel and Schiffer 2007, p. 61). With a merger discussion that began in 2007/2008 between SUEK, the largest Russian Coal producer, and Gazprom, Russia's largest company and gas producer, it seems rather likely that Russia will increase its own coal use in order to free up gas export capacity (see McCloskey 2007, 9 Feb 2007, Issue 153, 1). Today's known plans are such that Russia will increase the share of coal from around 23% of generation to about 30% in the future, with gas falling from around 66% to 58%. The result will be an increase in coal consumption of about 50 million tons per annum (Global Insight-World 2007, Section 2, p. 24). This development, if it materializes, will put further supply constraints on the global seaborne steam coal market.

3.4.5.1 Russia's Production and Exports

Russia produced 309 million tons of hard coal in 2006 (VDKI 2006; Ritschel and Schiffer 2007). Of this, 239 million tons were steam coal. In total, Russia exported 77 and 68 million tons of hard coal and steam coal, respectively. Another approximately 7 million tons were exported via the green border by rail. Production is centered in Kuzbass, Siberia, where about 80% of Russia's coal is produced. About two-third of Russia's production is opencast. The production is concentrated in such a way that the top eight producers account for almost 70% of Russia's coal production and over 80% of Russia's coal exports (see Table 3.10). Russia's coal production, unlike other industry sectors, is largely in private hands with no government involvement.

Unfortunately, there is no information available on export mine capacities. Thus, it is not possible to make statements about capacity utilization that are backed up by figures. Based on market knowledge, I can state that Russia's mines still have exportable coal, though they are running ever closer to capacity. Infrastructure, mostly rail, is the key bottleneck for the country.

Two-thirds, or just over 50 million tons, of Russian hard coal are shipped to the EU-15 countries. Another 11 million tons are shipped to Japan and 5 million tons to

Table 3.10 Russia's most important steam coal producers

Company	Output 2006 (million tons)	Exports 2006 (million tons)	Notes
SUEK	89	28	In merger talks with Gazprom
Kuzbassrazresugol (KRU)	41	18	Exports mostly managed by Glencore
Yuzhkuzbassugol	16	3	Private
South Kuzbass	17	5	Private
Russian Coal	14	2	Private
SeverStal Resource	13	1	Private
Sibuglemet	12	1	Private
Raspadskaya	10	6	Private
Total	**212**	**63**	
Percentage of total	69	82	
Russia 2006	309	77	

Source: Ritschel and Schiffer (2007), Deutsche UFG (2006), and Global Insight-Russia (2007, p. 30); Author's market interviews.

Korea. I expect that the Far Eastern sales will gain in relative importance, not just for infrastructure reasons (see below).

3.4.5.2 Russia's Coal Quality

Russian coal exports are of relatively high quality, with relatively high calorific values and low sulfur contents. Much of Russia's coal for export is improved through washing programs. However, it is difficult to estimate how much of the coal is truly washed. Because of high inland transportation costs, the producers will only sell high-quality coal to the market. The lower-quality coal, including about 70 million tons of produced lignite, is used domestically.

The difficult and lengthy transportation on old rail cars results in contamination of a lot of Russian coal with metal and other debris. The ports employ extensive metal cleaning technology, but many consumers still complain about problems with scrap and other contamination. In wintertime, Russian coal is at risk of freezing, especially when shipped from Murmansk. Therefore, despite Russia's low sulfur contents and good CVs, the quality of Russian coal is not considered the best because of the lack of quality homogeneity.

Coal from the Kuzbass region is known for its low sulfur contents. This is probably the only advantage of being far from the sea. Also, the reserves were far removed from saltwater when the coal developed millions of years ago. Because of the low sulfur contents, Russian coal is often used for blending with higher-sulfur coal in European power stations. Thus, despite the high FOB costs and contamination problems, I do not expect Russian coal to disappear from the international coal supply arena.

3.4.5.3 Russia's Infrastructure

Logistically speaking, Russian coal is very important for Eastern Europe, Turkey, and the Nordic countries (i.e., Finland and Denmark). Despite being a very high-cost producer, this is another key reason why Russia will continue to play an important role in the Atlantic market, especially for the Nordic countries, northern Germany, and the United Kingdom. It is important to note that ports that can only accept smaller vessels (such as many UK ports) often have no option but to take coal from nearby ports in countries such as Russia or Poland, unless they want to rely on transshipment via the ARA region. This factor becomes more important when freight prices are high.

Murmansk and Riga are the most important western ports for Russian coal, Yuzhny is the main port on the Black Sea, and Vostochny and Vanino are the main ports in the Far East. Russian coal needs to be transported by rail up to 4,000 km from Kuzbass to the ports. Thus, the key infrastructural constraint is rail capacity. The rail company, RhZD, is still 100% government-controlled (similar to the situation in South Africa). Rail tariffs have been progressively increasing from below US $10/ton in 2003 to an average of US $23/ton by end of 2006, and about US $35/ton by mid-2008 when this was written (see Global Insight-World 2007, Section 3, 23). Some insiders claim that the increased rail tariffs are one way for the government to recoup the money lost by the private producers' offshore marketing departments and low transfer pricing policies.

3.4.5.4 Russia's Costs

As mentioned above, Russia is – next to Poland – the highest marginal cost producer in the world seaborne traded steam coal market. Russia employs very sophisticated and high-quality production techniques that can easily compete with Australian and South African operations, but the logistical challenge cannot be overcome. Also, the ports in the Baltic countries started to ask for US $10/ton and more in early 2008 for transshipment of coal through their facilities.

Inland transportation costs are expected to rise even further in the future. I estimate the 2006 marginal FOB cost to be in the range of US $39–53/ton. Railage alone accounts for 62–72% of total FOB cost. Russian costs will continue to go up; by mid-2008, when this was written, FOB costs had already increased to US $65–70/ton and above. For the market this means that Russia is likely to be the first to reduce export volumes when international coal availability becomes abundant and sea freight rates decrease substantially. However, as mentioned above, the logistical advantage in Northern and Baltic Europe and the low sulfur advantage strongly support continued long-term Russian coal exports.

3.4.6 China

I will discuss China a bit more briefly than the previous exporting nations. This is not because China is not important; on the contrary. The reason is that, in the future,

China's role will likely be less that of a coal-supplying country to the world market; China became a net importer for the first time in its history for a few months in 2007. I will rather focus on the importance of China as a swing supplier to the global seaborne steam coal market. China is also probably the most important determinant in what will happen to the future of this market.

3.4.6.1 Basic Background

China is a nation of large numbers. I summarize some key facts relevant to the evaluation of China's role in the coal market below. The following sources were used: Ritschel and Schiffer (2007), BGR (2006), BGR-China (2007), and IEA-Statistics (2005).

- China consumes 27% of world coal production used in power generation. Yet, with 1.3 billion inhabitants, China still only consumes 2,200 KWh electricity per capita compared to 6,400 KWh in Germany.
- The country relies to 79% on coal for its electricity generation.
- China produced 2.4 billion tons of coal in 2006 (almost 45% of global coal production), with more than 50% of this coal utilized in the power industry.
- The top eight producers account for 590 million tons, or 25%, of China's production.
- Chinese coal reserves and resources are ample, translating into a remaining production potential of over 1,800 years.
- Exports have historically been to Japan, South Korea, and Taiwan, in that order. Thus, the Pacific market is strongly influenced by China's coal policy.
- The government has started to adopt a series of measures resulting in ongoing consolidation. The goal is, for example, to close 7,000 smaller pits and to form 5–7 large coal companies with an annual output of over 100 million tons.
- China's 2006 marginal FOB costs are in the range of US $33–45/ton; costs will rise further for geological and logistical reasons. However, China's port infrastructure is sufficient.

3.4.6.2 China and the World Coal Market

The key consumers of Chinese coal – Japan, South Korea, and Taiwan – are increasingly looking for replacements from Australia, Indonesia, Russia, and even South Africa. The reason for this is that China is on the verge of becoming a net importer and thus these countries cannot rely on China in the long run. However, many trading firms and market analysts forget that China's cost structure and infrastructure are such that large amounts of exports are possible without much prior notice. In the medium term, everything points toward China becoming a net importer. However, GDP growth, new technology, and world market prices may at any time result in China becoming a significant if not one of the largest global coal exporters for short periods of time. With 2.4 billion tons of production, a theoretical 100 million tons of added exports only account for 4% of China's production but would account for 17% of 2006 global steam coal exports. The same risk of course goes the other way: if China imported 100 million more tons of coal, this would shake the global market and have a large impact on the 'old' economies in Europe, with the possibility of very large price increases as seen during 2008.

This risk of China easily becoming a swing supplier is a key reason for increased price volatility in the international coal arena. The various market participants can prepare themselves for the 'China Risk Factor' in a number of ways as detailed in Table 3.11.

Table 3.11 How to prepare for the 'China risk factor'

Consumers
– Increase the percentage of procurement that is long-term contracted (which, the way, this is against the current international trend) to fight volatility
– Forge long-term relationships with key producers and traders to ensure contract performance even in 'bad' times. Keep your eyes open toward Chinese exports
– Keep or make your power plants as flexible as possible in terms of acceptable coal quality. This becomes especially important in times of coal scarcity
Producers
– Keep your marginal cost as low as possible and don't rely on high coal prices forever. Production costs below comparable Chinese cost levels will ensure long-term profitability
– Be flexible and keep your eyes open toward exporting to China if demand exists
Traders
– Volatility caused by China means increased opportunity but also increased performance risk from both consumers and producers, especially for traders. Thus, forge long-term relationships with key consumers and producers and diversify
– If you are active in the Pacific market you should have a good Chinese partner or your own Chinese office to keep you abreast of developments. Either way, China will always export and import at the same time, even if China becomes a net importer

Source: Author's market analysis.

3.4.7 Colombia and Venezuela

Colombia is the key exporting country in Latin America with 2006 exports reaching 61 million tons (VDKI 2006). I include Venezuela (9 million tons of exports in 2006) in this section since the two countries are very similar in terms of coal qualities and geographical location. In fact, some Colombian coal is also shipped via Venezuelan ports. The United States consumes most of Colombia's and Venezuela's coal. In total, almost 40%, or about 27 million tons, of the 70 million exported tons was delivered to the United States. Europe (EU-15) consumes as much as the United States; thus these two markets comprise close to 80% of the region's exports.

The Colombian supply market is considered a duopoly, with Cerrejon, a BHPB/Anglo/Xstrata-Glencore consortium (28 million tons of exports in 2006) and US-based Drummond (21 million tons of exports in 2006) accounting for almost 80% of export volumes. Adding Prodeco (Carbones del la Jagua, Glencore) with 8 million tons of exports, the top three producers emerge with a 93% export market share (VDKI 2006; Ritschel and Schiffer 2007). Production in these countries is highly efficient and I estimate that marginal 2006 FOB costs range from US $23/ton for the high-volume mines to US $51/ton for the smaller mines far away from the ports. Some newcomers have been granted mining licenses.

Given the importance of oil for Venezuela, coal will retain a small niche presence there. However, there is potential for 20+ million tons per annum in exports if the ports are expanded, which Vale (former CVRD) is now working on. Colombia's coal generally has a very low ash content, which results in above-average calorific values. When buying some of the very high-CV Colombian coals on FOB terms, the buyer will benefit from reduced relative transportation costs and efficient coal burning. The drawbacks of the coal are – generally speaking – relatively high moisture contents, low HGI, thus hard coal and, in some cases, a proneness to self-ignition. However, this type of coal may be used as PCI coal for the steel industry (used to replace coking coal).

The future will see rising export volumes from Colombia and Venezuela. Reserves are large enough and various projects to improve rail and port infrastructure are under way. These countries do not yet consume much coal themselves and even if that consumption should increase, it would do so from a small basis. Demand in North, Central, and South America is increasing and Colombia will remain a key supplier for these American markets as well as Europe. Colombia does not need to fear rising freight rates as much as, for instance, South Africa, as there is sufficient demand in nearby regions.

3.4.8 USA, Vietnam, Poland, and Canada

3.4.8.1 United States

With 1.1 billion tons of production in 2006 the United States was the world's second-largest coal producer after China. Net seaborne exports totaled 26 million tons in 2006. Another 20 million tons were exported to Canada via the green border. The four biggest exporting companies (Peabody, Consol, Massey, and Alpha) accounted for about 90% of 2006 exports (Ritschel and Schiffer 2007; Kopal 2007; Global Insight-World 2007).

The United States, like China, is a swing supplier, but to a much lesser extent. In the second half of 2007, for instance, exports increased more substantially because of the increased international coal price levels. Demand in the United States is strong but may decline and infrastructure (rail and ports) is currently at its limit. There are also strong political and public movements against coal power inside the United States that is likely to lower indigenous coal demand (PESD 2009).

I expect the United States to continue playing the role it has: supplying the world market, notably Europe, whenever it is opportune and the logistics allow it. Nevertheless, US coal tends to be of high sulfur content, thus usually commanding a somewhat lower price than, for instance, South African coal. US coal has no importance for the Pacific market. Since the Pacific is where the major demand increases originate, US coal may, to some extent, replace South African and Russian coal, which will increasingly be exported to Asia.

3.4.8.2 Vietnam

In the coal industry, Vietnam is known for its high-quality anthracite products. Exports in 2006 exports totaled 30 million tons (44 million tons production), of which anthracite probably accounted for more than 20 million tons (Ritschel and Schiffer 2007; Kopal 2007; Global Insight-World 2007). Vietnam plays a rather small role in the steam coal market. In fact, as the domestic economy grows the government is supporting programs to further limit exports. Nevertheless, local production will increase further over time. Because of geographical proximity, Vietnam's coal is well suited to the southern Chinese market. Even Vietnamese anthracite tends to compete in a rather local market and is of little significance for the world market. Some Indonesian coal has been exported to Vietnam for their paper industry. I expect that Vietnam will increase its imports over time and become another valuable client for Indonesia.

3.4.8.3 Poland

Poland produced 96 million tons of coal in 2006. Historically, Poland has exported significant amounts of coal to Germany and northern Europe, and even sometimes to the Mediterranean countries. Seaborne exports reached 7 million tons in 2006 (total 16 million tons including green border trade), down from almost 11 million tons in 2005 (total close to 20 million tons including green border trade). In 2007, Poland reduced its exports further (Ritschel and Schiffer 2007; Kopal 2007; Global Insight-World 2007). For Germany, Poland remains an important supplier via the green border. Vattenfall's Berlin power stations in particular rely heavily on Polish coal. However, the company is looking at alternative supplies via the North and Baltic Seas.

Poland is the most expensive producer on an FOB marginal cost basis. Therefore the country plays a role in setting the floor price in Europe. I expect marginal 2006 FOB costs to be around US $71/ton. This figure will continue to increase as Poland struggles with EU legal requirements. State-owned mines are often unaware of their true cost structure. Some efficient mines have been closed down but may be reopened in the future. Poland is in fact becoming an importing country, especially of Russian coal, via its ports in the Baltic and via its borders with Ukraine and Belarus.

3.4.8.4 Canada

Canada produced 70 million tons of hard coal in 2006, of which 40 million tons was coking coal and PCI coal. The country exported 28 million tons, but only 3 million tons of steam coal (Ritschel and Schiffer 2007; Kopal 2007; Global Insight-World 2007). Thus, Canada plays a major role in the coking coal market, but not in the steam coal market, and therefore is of little relevance for this study.

On a final note about the fringe suppliers discussed above, I would like to point out that, economically speaking, these smaller supplying countries have an interest in the main supplying countries capturing monopoly rent or exercising their market power. Kolstad and Abbey (1983) have shown that exercising market power can increase the market share for competitive fringe suppliers. This is an interesting realization that may be one factor influencing the coal supply market's gradual move away from perfect competition.

3.5 Competitive Conditions of Coal Supply

Section 3.4 described the various countries that supply steam coal to the international coal trade. In this section I will briefly discuss the overall concentration of the steam coal market. The coal supply market is becoming more concentrated on a global level, and even more so in the various supply regions. Looking at the entire world, Table 3.12 shows that the 17 top coal producers account for almost 55% of world hard coal exports and one-third of world production. Taking that same table and analyzing it for world export share we find that the top 5 companies (BHP, Xstrata, Anglo American, Adaro, and Rio Tinto) account for close to one-third of global exports and that the top 10 companies already account for 45% of global hard coal export volume.

On a regional level, the concentration is much stronger. Over 60% of European (2006) imports come from four major producers and almost 40% of Pacific imports come from four major producers. Buisson (2007) analyzed the various supply markets and discovered that five major producers – BHP-Billiton, Anglo American, Xstrata/Glencore, Rio Tinto, and Drummond (Atlantic market only) – together largely control the South African (86%) and Colombian (82%) markets. Australian (67%), Russian (40%), and Indonesian (38%) exports are also very much influenced by these five players. From my own market analysis, I can state that about every fourth ton of steam coal traded passes through Glencore/Xstrata's hands. That means that the private Swiss-based Glencore group founded by Mark Rich and partners handles 150 million tons or more. (For a list of the 16 most influential coal market players according to the Coal Market Survey, please refer to Appendix D.)

Figure 3.9 graphically depicts the points discussed above. Such a high concentration of coal exporters in the various markets has an effect especially in times of coal scarcity, such as in the years 2007/2008. I will discuss pricing in more detail in Section 5.3. Overall, it is expected that consolidation will continue. Almost 25%

Table 3.12 Top world hard coal producers

No.	Company	Resident	Ownership	2006 production (million tons)	2006 exports (million tons)
1	Coal India	India	Government	343	0
2	Peabody	South Africa	57% Lehman	232	14
3	Shenhua	China	Government	203	26
4	Rio Tinto	Australia	Public	154	31
5	Arch	United States	Public	127	0
6	Anglo American	United Kingdom	Public	98	42
7	China Coal	China	Government	91	27
8	Suek	Russia	Private	90	28
9	BHP Billiton	Australia	Public	86	61
10	Xstrata	Switzerland	Public/Glencore	77	53
11	Consol	United States	Public	67	10
12	Bumi Resources	Indonesia	Private	51	24
13	Kru	Russia	Private	44	18
14	Massey	United States	Private	35	6
15	Adarao	Indonesia	Private	34	34
16	Cerrejon	Colombia	Private/Xstrata	28	27
17	Drummond	United States	Private	21	21
Total				**1,781**	**422**
Share of world hard coal (%)				33	54
World hard coal (million tons)				**5,400**	**782**

Source: McCloskey (2007), and Ritschel and Schiffer (2007), Author's market research and analysis.

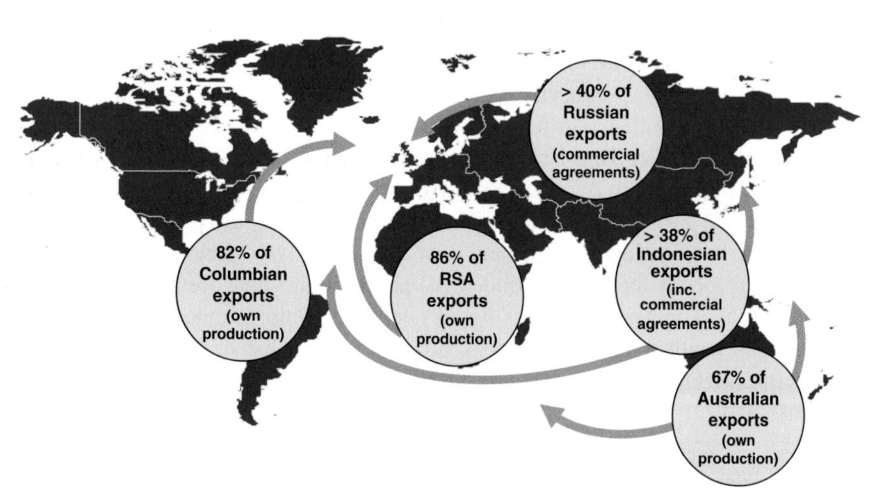

Fig. 3.9 Presence of top five coal exporters in 2006 (Note: Top five – BHP, Anglo, Xstrata/Glencore, Rio Tinto, and Drummond. Source: Buisson 2007; Author's market analysis)

of the Coal Market Survey participants ($n = 167$, 33 'likely,' 5 'almost certain') answered that the development of a 'Coal-PEC' is likely or certain within the next 5–10 years. (For more details please refer to Appendix D.) However, some markets, such as Russia and Indonesia, will remain more fragmented.

3.6 Coal Mining/Production

This section looks at the coal mining process. Understanding mining is important for this economic study because coal mining costs are a key component of the marginal FOB costs relevant for global long-term pricing. Based on my research and development of the 2006 marginal cost curve for global steam coal supply to the international markets, I estimate the average cost breakdown as detailed in Fig. 3.10.

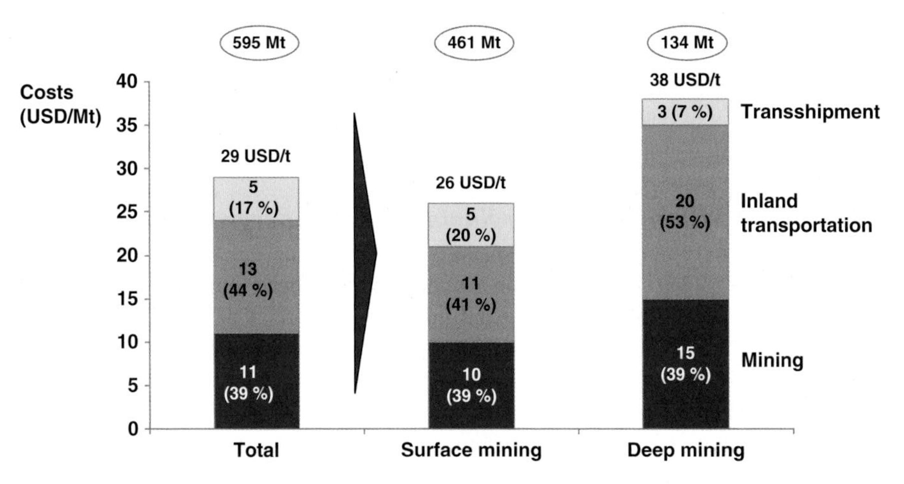

Fig. 3.10 2006 Global average marginal FOB costs (Source: Author's market interviews and market analysis; FOB cost analysis)

Mining costs account for almost 40% of the total marginal FOB costs in 2006, inland transportation accounts for 44%, and transshipment for 17%. However, the mining companies have relatively little influence over inland transportation and transshipment costs.

3.6.1 Optimal Use of Resources

Before discussing specific details of coal production, I will briefly summarize the economic implications of utilizing any nonrenewable natural reserve. When mining companies own a reserve they have to make a decision: (a) to leave the reserve in the ground until market prices rise further due to resource scarcity; or (b) to start mining the reserve now and invest the resulting profits in the capital market.

Erdmann and Zweifel (2008, pp. 127–129) summarize the economic theory behind this decision-making process, which was first developed by Harold Hotelling (1931). Key assumptions are as follows:

- Efficient raw material and capital markets;
- Owners of reserves maximize profits with full knowledge about reserves and mining costs;
- No storage of mined reserves; and
- Constant demand elasticity and constant price.

Despite these assumptions, Hotelling (1931) came to the conclusion that the price of a raw material will increase, even assuming constant marginal cost of production. I will make the point in Chapter 6 that the marginal costs of raw material production are in fact not constant but increasing. This increase results from the increased scarcity of the natural reserve. The scarcity rent corresponds to the economic value of the reserve in the ground. In efficient markets, one pays this scarcity rent when acquiring a natural resource. In later chapters I will discuss scarcity and the implication on pricing for steam coal. But first, in the following sections, I will discuss specific aspects of coal mining.

3.6.2 *Coal Mining Methods: Surface and Underground Mining*

About 60% of global coal production is underground or deep mining and the remaining 40% is surface mining (World Coal Institute-Resource Coal 2005). Because of its inherent cost advantage, surface mining is much more important for export coal mines, with an average share of about three-quarters and only about one-quarter extracted through underground mining (Author's FOB cost analysis). The choice of coal mining method is largely determined by the geological location of the coal seams. Typically, larger coal seams at a depth below 150–200 m are mined in underground mines (though, this also varies from country to country). Figure 3.11 summarizes the stages of a typical mining project. The mining company decides whether the resource can be surface-mined by the first stage of the exploration phase at the latest.

The methods of mining mentioned above differ not only in terms of marginal costs (export marginal mining costs of US $15/ton for deep mining vs. US $11/ton for surface mining) but also in terms of investment costs. However, since only the marginal costs are relevant for the long-term competitiveness of a product, surface mining will always have an advantage. This is borne out by the prevalence of surface mining in the global seaborne steam coal market.

3.6.2.1 Underground Mining

Below I will focus on the coal extraction process, which is part of the exploitation phase of a mining project (see Fig. 3.11). Underground mining is usually conducted in one of the following two ways:

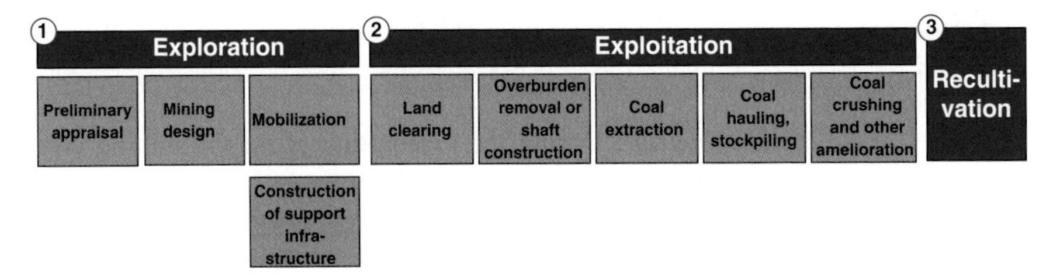

Fig. 3.11 Process of a typical mining project (Source: Author's research and analysis)

1. long wall mining and
2. room and pillar mining.

The long wall mining method is much more efficient and used in the world's largest mines (i.e., in Russia, Australia, and the United States). Here a mechanical extraction machine or *shearers* extracts entire seams of coal. The coal face from which the machines extract can vary in length from 100 to 400 m. Semiautomated (self-advancing) equipment supports the roof of the coal seams temporarily with hydraulically powered supports. In most cases, the coal roof collapses once the machine has moved on. These machines can extract from various seam thicknesses, from between 3 m and 5+ m. Seeing such equipment in operation is very impressive, but the mining company often has to spend tens of millions of US dollars just to procure one of these machines. Also, the planning, installation, and training times involved are very long and costly.

Here lies the advantage of room and pillar mining. Mobile machinery that costs US $1–5 million per mine can be used and set up very quickly. Here only 50–70% of the coal is mined in each seam; the remaining 30–50% of the coal remains in the mine and forms the pillars that support the roof. One moves through the seam from left to right, leaving some coal unmined to form the pillars. The resulting pattern, when viewed from above, looks a bit like a chess board.

3.6.2.2 Surface Mining

Surface mining is the oldest way of extracting coal from the ground. At its core is the very basic 'shovel and truck' exploitation system (for more details, see Fig. 3.11). One differentiates between continuous and discontinuous surface mining. Continuous mines often employ large wheel bucket excavators, reclaimers, and extensive systems of conveyor belts to move the overburden and coal. Discontinuous mining mostly uses excavators, dozers, and trucks to extract and move overburden and coal. The basic surface mining exploitation system is accomplished in five stages: (1) the selected land area is cleared; (2) the overburden is removed; (3) the coal is extracted; (4) the coal is trucked to the intermediary stock piles at the mine; and (5) the coal is crushed or otherwise processed.

Stage 1. Land clearing: Depending on the size of the mine area and its location, this process may be as simple as removing trees and fauna from the land. In larger and more populated mine areas, for instance the large lignite mines in the Lausitz region in the east of Germany, this process may also involve relocating villages, houses, and roads.

Stage 2. Overburden removal: During the exploration phase the mining company determined the mining plan and the size of the reserve. During this process the so-called Overburden Ratio or 'OB ratio' (the average number of cubic meters of overburden that needs to be removed in order to extract one ton of coal) is determined. The overburden is removed either in a continuous or in a discontinuous way, as explained above. The overburden needs to be replaced strategically so it does not have to be moved twice. Sophisticated mines use the overburden they remove from the start of the mine to refill the mining area after the coal has been extracted at the end of the mine. The OB ratio is the key economic factor determining mining cost, because removing one cubic meter of sand costs exactly as much as removing/extracting one cubic meter of coal (depending on its quality, one ton of raw coal equals usually between 0.9 and 1.3 cubic meters of sand)

Stage 3. Coal extraction: The coal is extracted either in a continuous or – more often in the case of hard coal – in a discontinuous way, the same as the overburden. The key economic factor during coal extraction is to leave as little coal as possible unmined. In fact, some countries have regulations that determine how much coal can remain unmined in the ground. In the case of Russia, usually a maximum of 20% of the reserve may remain unmined, otherwise the mining company risks losing its exploitation licenses. With surface mining, most of the coal – usually 85–95% of the reserve – can be extracted. With underground mining, depending on the method, this figure is usually lower. The risk of extracting too high a percentage of the reserve is that the coal quality will decline for two reasons: (1) the coal closer to the roof of the seam often has higher sulfur contents or other impurities, because geologically the upper parts filter the water; and (2) the closer you come to the edge of the seam, the higher the chance of extracting sand and other sediments including the ash content of the coal.

Stage 4. Coal hauling, stockpiling: A number of very large trucks or extensive conveyor belt systems move the coal from the extraction site to the intermediary stockpiles. The key economic factor is to design a road and conveyor belt system that is inexpensive to maintain but minimizes distances. Especially with rising fuel costs, the design phase is therefore gaining importance. Smaller and less-efficient mines will lose more and more competitive advantage in times of high fuel costs. An added layer of complexity is caused by the fact that the mine moves, thus requiring roads and any conveyor belt systems to move as well.

Stage 5. Coal crushing and other amelioration: The mined coal is extracted in what we call 'raw' or ROM (run-of-mine) form. All ROM coal needs to be treated in one or more ways, including but not limited to (a) crushing, (b) screening, (c) washing, and (d) drying. For surface mining and for power plant use, all coal is crushed and often screened. In South Africa and much of Russia, exported coal is also washed. Section 3.6.4 will discuss in further detail the coal amelioration processes.

3.6.3 Coal Mining Investments and Variable Operational Costs

In the following two sections I will analyze the investment and variable operations costs involved in coal mining. The variable operations costs in particular are important in determining market price levels. Investments costs will influence market price levels in times of scarcity. The basic economic theory behind natural reserve utilization was already discussed in Section 3.6.1.

3.6.3.1 Investments in Coal Mining

Investment costs in coal mines can be significant and need to be analyzed in order to understand the future economics of the industry. Especially in scarce times, even the most expensive producer in the world will want to recoup his investment costs (for more detail on economic theory about this subject, please refer to Chapters 5 and 6 and Appendices E and G). Investment costs include expenditure on developing the deposits before or during actual production. This will include the following:

- purchase costs for land or mining rights;
- one-time license costs; and
- prospecting and exploration costs.

It has been estimated and confirmed by Kopal (2007) and Ritschel and Schiffer (2007) that total investment costs average about US $60 for each annual production ton of coal mining capacity (see Fig. 3.12). Thus, when one develops a mine with one million ton of output per annum, one will pay on average US $60 million upfront. Total financial costs are currently estimated at US $6–7 per annual production ton, consisting of US $2.5–3/ton depreciation assuming a 20-year

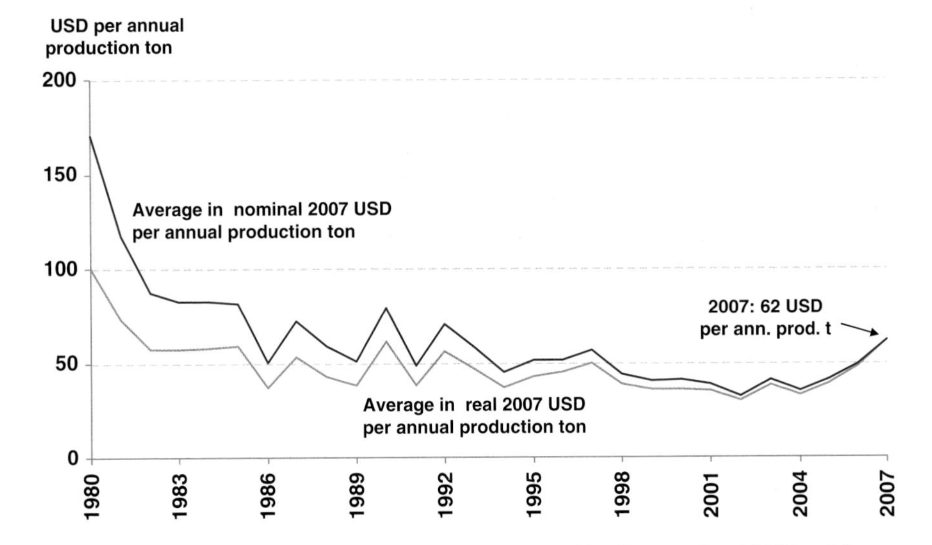

Fig. 3.12 Specific investment cost in export mining, 1980–2007 (Source: Kopal 2007, p. 16)

lifetime of a mine and US $3.5–4/ton interest (assuming a 10% average debt service rate).

Investment costs vary widely. For mines with little to no existing infrastructure nearby (i.e., lacking roads, rail tracks, water, energy supply, accommodation), this cost can be substantially higher. Some greenfield coal projects (requiring large-scale investment in infrastructure) in remote and undeveloped areas may require specific investments of up to US $160 for each annual production ton. From my own experience, however, I can also state that some smaller coal mines in logistically superior locations can be developed with investments as low as US $3–10 per annual production ton. This is especially true for small mines in Colombia, South Africa, and, of course, Indonesia.

It is interesting to see in Fig. 3.12 that real investment costs declined from the early 1990s, but have been increasing since 2004. The recent rise is not surprising, given the increased demand for mining equipment and increased raw materials costs in general (please refer to Fig. 2.3 on page 19 and Fig. 2.4 on page 20 for more detail on price developments).

Relatively speaking, coal has received a very small share of global investments in energy. This is astonishing, given the importance of coal. As discussed previously (see Fig. 3.1) 25% of world primary energy and 40% of world electricity generation are based on coal. However, as detailed by the IEA World Energy Investment Outlook (2003) and summarized in Fig. 3.13, coal is projected to receive only 5% of global cumulative investments in fossil fuels between 2001 and 2030.

Kopal (2007) and Ritschel and Schiffer (2007) pointed out that this trend has changed recently. Coal will continue to increase its investment share compared to

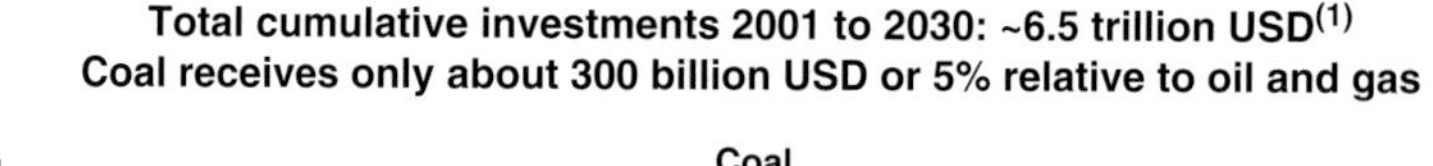

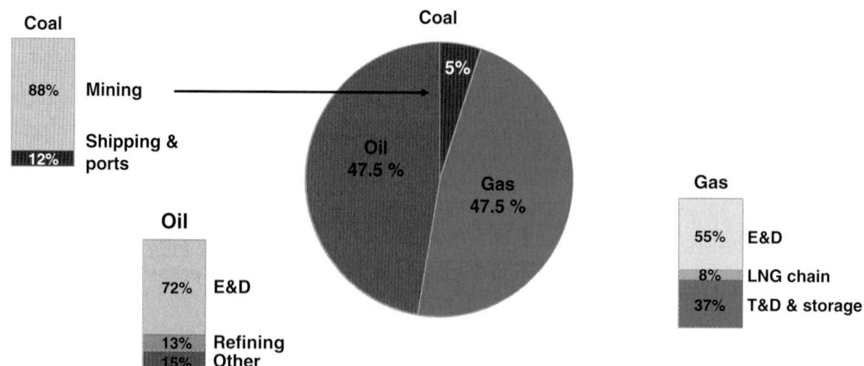

Fig. 3.13 World energy investments cumulative, 2001–2030 ([1] Global cumulative investment in energy: US $16 trillion, of which 60%, or US $9.5 trillion, invested in electricity generation, and US $6.5 trillion invested in energy resources. Note: T&D = transmission and distribution; E&D = exploration and development. Source: IEA World Energy Investment Outlook 2003, Author's research & analysis)

gas and oil. When coal increases its global investment share, this will translate into higher investment costs per reserve ton, which in turn will result in higher prices, especially in scarce times, such as recently in 2007/2008 (Fig. 3.13).

Nevertheless, the IEA has also determined and confirmed that coal is a less capital-intensive energy resource than oil or gas and has much room to remain competitive even when investment costs increase (IEA World Energy Investment Outlook 2003). In fact, gas requires 4.5 times and oil 5.8 times as much investment as coal, all calorific value adjusted. Therefore, coal carries less investment risk than gas and oil. When we take the accumulated investments throughout the entire supply chains of coal, oil, and gas from 2001 to 2030 and divide these by the respective accumulated production growth, we can see the relative investment requirements of each fossil fuel source:

- Coal: US $3.4/tce of production growth.
- Oil: US $15.4/tce of production growth.
- Gas: US $19.6/tce of production growth.

Today, it is crucial for mining companies to secure coal assets. Asset prices will increase further and coal will demand a 'fairer' share of its value. Deutsche Bank and other financial institutions usually calculate US $1 per reserve ton when valuing coal reserve assets. With profit margins increasing over time, this value is likely to rise in the future. This will lead to more funds entering the coal mining sector, which in turn will increase asset prices.

3.6.3.2 Coal Mining Operating Costs

Variable or marginal coal mining costs depend on a number of factors: (1) the type of mining operation (i.e., opencast vs. underground, continuous vs. discontinuous); (2) topography and available infrastructure (i.e., access to electricity and how hilly the region is); (3) the type of coal (i.e., steam coal vs. coking coal); (4) labor costs; and (5) productivity per man-year. The key components of average opencast mining costs are summarized in Table 3.13.

Table 3.13 Percentage breakdown of variable mining costs

Category	Typical share of total variable cost (opencast mine in developing world) (%)
Fuel	30–40
Maintenance and repair (incl. tires)	30–40
Labor	10–20
Other support functions and royalties	10–20

Source: Author's research and analysis.

It is interesting to see that about two-thirds to three-quarters of the variable costs are driven by fuel and maintenance and repair. Thus, the kind of equipment used in any opencast mine will determine how efficient the mine is. The demand for high-quality trucks and excavators vastly outstripped supply in the boom years of 2006–2008. Leasing and owning costs also increased substantially. Tires are out of stock and OEM spare parts are scarce. Today, miners and contractors struggle to procure the good-quality machinery and equipment they need in order to keep up with expansion plans. At the same time, fuel prices more than doubled in less than 12 months from 2007 to 2008. This was not surprising given the oil price increase in the same period (Fig. 3.14).

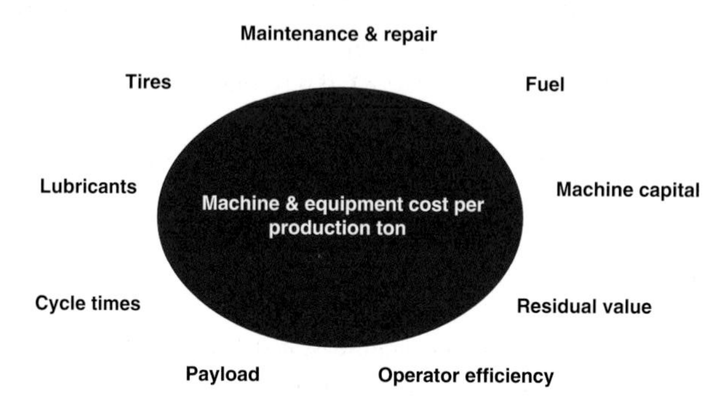

Fig. 3.14 Key drivers of efficient coal mining operations (Source: Author's research and analysis; Latham 2008)

The outlook for operating costs is not positive. There is substantial rationalization and productivity improvement potential in many smaller and medium-sized mining operations. However, the key components fuel and labor will likely increase further in cost. Coupled with higher investment and thus financing costs, it can be expected that coal production as a whole will become significantly more expensive in the future. The floor for coal prices on the international seaborne coal market will therefore increase further. Key reasons for this price increase were already discussed in Section 3.4.2. Here again is a brief summary of the drivers causing coal production costs to increase:

- new deposits require longer inland logistics;
- overburden ratios increase;
- fuel costs increase;
- machinery costs increase;
- royalties and other government charges are likely to increase; and
- coal qualities decrease.

3.6.4 Coal Amelioration and Coal Washing

As discussed in Section 3.6.2.2, crushing, screening, washing, and drying are typical ways of treating or ameliorating run-of-mine (ROM) coal after it has been extracted. Coal amelioration costs are an integral part of the total coal mining costs. It is important to note that all costs for the entire mining and logistics chain are paid on a per-ton basis. Thus, the earlier the coal is ameliorated, the better, since all costs (usually only logistics-related) are then lower per calorific unit. For instance, let's assume that a South African producer extracts and hauls the coal at a marginal cost of US $8/ton. The producer then washes the coal at a cost of US $3/feed ton to reduce the ash content, producing a yield of 65%. Thus, for each ton of ROM coal the producer gets 0.65 tons of washed product, yielding 6,200 kcal/kg NAR. The coal is then transported to the port and transshipped at a total marginal cost of US $30/ton. The coal is finally sold for US $60/ton. The sales basis of 6,000 kcal/kg NAR is the international standard for South African coal. The profit achieved for this coal thus equals the sales price minus the true marginal FOB cost.

Washing means treating coal in order to reduce the ash content. Before washing, the coal is usually crushed and screened to remove the very fine material (for instance, 2 mm in size) because fine material does not wash very well and can clog the machinery. The coarser material is treated using 'dense medium separation.' Here the coal is separated from other impurities by being floated in a tank containing a liquid of specific gravity. Usually, this liquid is a suspension of finely ground magnetite. The lighter coal floats and can be separated from the heavier rock and other impure materials that sink (World Coal Institute-Resource Coal 2005).

Figure 3.15 shows a mobile wash plant capable of treating 100 tons of ROM coal per hour. Larger stationary wash plants are capable of treating 300+ tons of ROM material per hour. Assuming 20-h operation for 25 days a month over 12 months, you can process 600,000 tons of ROM coal per year for each 100 tons/h washing capacity.

Fig. 3.15 Mobile wash plant with 100 tons/h feed capacity in South Africa (Source: Author's photograph)

When washing coal, the yield is the most important factor. The more ash you wash out, the lower is the yield and the more the discard – the name of the washing by-product – you receive. Thus, relative costs increase and the need to find an outlet for the discard can also reduce your economic return. Therefore, mining companies take into consideration the increased sales price for higher calorific value material, the cost of washing, the yield, and the proceeds, if any, from discard material.

3.6.5 Environmental and Safety Issues Associated with Coal Production

There are a number of environmental and safety issues that need to be considered when preparing a mining operation. In this section only environmental issues associated with the production of coal will be discussed; issues arising from the burning of coal will be discussed in Section 4.5. I will also discuss methane emissions and other environmental concerns.

3.6.5.1 The Environment and Coal Production

As with any industrial process, the production of coal strains the environment. State-of-the-art mine planning can minimize the environmental side effects. The effect on the environment can be classified in four categories (see World Coal Institute-Secure Energy 2005):

- emissions from fuel-consuming equipment;
- land disturbance and mine subsidence;
- water, dust and noise pollution; and
- methane emissions.

Emissions from Equipment

IEA-Oil (2007) estimates a total 2005 demand for motor gasoline and middle distillates of around 2.1 billion tons (compared to total oil product demand of around 3.9 billion tons). As discussed in Section 3.6.3.2, fuel accounts for about one-third of the operational costs in a typical noncontinuous opencast mine. What this means in effect is that each ton of noncontinuous opencast coal requires about 2–3 liters of fuel. Therefore, about 250 million liters of fuel are required to mine 100 million tons of noncontinuous opencast coal. Thus, coal mining results in the emission of significant amounts of carbon dioxide due to fuel consumption, which is not accounted for when considering carbon dioxide emissions from coal combustion alone, as most statistics do.

Land Disturbance and Mine Subsidence

Opencast mining requires land. During the mining process, vegetation, animal life, infrastructure, and even housing can be affected. Some older underground mines are known to have caused land subsidence. Especially when mining beneath inhabited areas, modern mine planners carefully calculate how much coal can be taken out and how the roof has to be supported.

Water, Dust, and Noise Pollution

During mining a chemical reaction between water and rocks containing sulfur-bearing minerals can result in acid mine drainage (AMD), a metal-rich water. AMD is formed when the pyrite reacts with air and water to form sulfuric acid and dissolved iron. This acid runoff dissolves heavy metals such as copper, lead, and mercury that are emitted into the ground (World Coal Institute-Secure Energy 2005). AMD can be minimized using water treatment plants and effective mine planning. Water can also be polluted during coal washing, which requires large amounts of water. Wash plant licenses are therefore only granted with strict environmental requirements. Dust and noise pollution can be a problem when coal is surface-mined near inhabited areas. Also, mine workers are subject to dust and noise pollution, and safety measures need to be installed in such mine operations.

Methane Emissions

Methane emissions can occur during underground mining. Methane is 21 times more harmful to the planet (in terms of global warming) than carbon dioxide. Methane needs to be vented during underground mining in order to reduce the risk of methane explosions. Frondel et al. (2007) and Steenblik and Coronyannakis (1995) point out that about 15 tons of methane are emitted for each ton of coal production. They argue, therefore, that a number of older underground mines, especially in Europe, should be closed down immediately in order to substantially reduce greenhouse gas emissions. Three million tons of coal mined underground in Europe produces approximately one million tons of CO_2 equivalent.

3.6.5.2 Safety in Coal Production

Larger mine accidents are regularly reported in the media. Many of these accidents occur in older mines in Eastern Europe and Asia. In 2004 *China Daily* reported that 80% of all deaths in coal mining accidents – according to statistics compiled by the Chinese State Administration of Work Safety (SAWS) – occur in China. However, mining operations in Australia, the United States, and other countries also experience underground mine accidents with collapsing roofs or fumigation from time to time.

Modern mines rarely experience safety problems. Effective mine planning can greatly reduce and avoid safety hazards. Existing mines need to be reinvestigated and international treaties should be signed to make coal mining safer. International mining conglomerates have done a lot but need to do more to make mining safer.

Another factor to be considered with mining safety is the employment of inexpensive and often untrained labor in some of the world's underground mines. Today, however, most countries have laws regulating the hiring of untrained and trained personnel in mines.

Having discussed in Chapter 3 the sources of coal and coal as a resource, thus analyzing supply, I now turn to the use of coal, thus analyzing demand, in Chapter 4.

Chapter 4
Use of Coal – Power Generation and More

4.1 Introduction

Electricity is one of the most important elements of human life today. However, about 2.4 billion people still rely on primitive biomass fuels to meet their household energy and heat requirements. According to the World Health Organization (WHO) 1.6 million people die each year from the effects of burning solid fuels indoors. Dependable and affordable access to electricity is essential for improving people's health, providing education and information services, improving living conditions, and freeing up time from the gathering of fuel. Electrification of rural areas and energy consumption growth are therefore not only environmental issues but also a necessity for the human race. Reliable access to electricity will allow (1) more people to work in agriculture and manufacturing jobs, freeing them from unproductive fuel-collecting tasks; (2) better use of agricultural or protected natural land; (3) improvement of public health; and (4) the use of modern appliances and lighting, which improves productivity. In short, electrification spurs a nation's GDP development (World Coal Institute – Resource Coal 2005, p. 21).

Our ancestors tamed fire for the first time 300,000–800,000 years ago. Since then, light and warmth from burning biomass has fueled – in the true sense of the word – the development of humankind. It took many thousands of years before humans discovered the advantages of burning coal rather than wood. Historians believe that the Chinese were at the forefront of this development. There are some reports that the Chinese were already using coal about 3,000 years ago for casting coins and smelting metal products. It took about 500 years longer for Europeans to utilize coal for energy generations. The Greeks and Romans were first. In the Middle Ages, the use of coal was widespread and the first trading of coal apparently occurred between England and Belgium. The industrial revolution in the eighteenth and nineteenth centuries would not have been possible without coal. The steam engine, powered by coal, was invented by James Watt in 1769. In the nineteenth century, many cities – starting with London – produced so-called town gas from coal to light their city streets using the coal gasification process (World Coal Institute – Resource Coal 2005, p. 19).

Today, coal is used in a variety of applications: (1) power generation; (2) steel production; (3) coal liquefaction; (4) cement production; and (5) other

L. Schernikau, *Economics of the International Coal Trade*,
DOI 10.1007/978-90-481-9240-3_4,

Table 4.1 Use of world hard coal by sector, 1980–2006

	1980		2006	
Consumption	Bln tons	%	Bln tons	%
Total hard coal consumed, of which	2.8		5.4	
–Power plants	1.0	36	4.0	74
–Steel industry	0.6	21	0.7	13
–Heat market, cement, and other	1.2	43	0.7	13

Sources: Author's research & analysis; VDKI (2006), Ritschel and Schiffer (2007), IEA Energy Outlook (2007)

applications including but not limited to household consumption, alumina refineries, paper manufacturers, chemical industry (such as for soda production), the pharmaceutical industry, and for specialist products (such as activated carbon, carbon fiber, and silicon metal).

However, almost three-quarters of the 5.4 billion tons of coal produced worldwide is used for power generation. The remainder is used for steel production (13%) and other applications (13%). Table 4.1 summarizes the uses of all produced hard coal. Please note that 4.7 billion tons of the total 5.4 billion tons of hard coal produced in 2006 was steam coal. Since 4 billion tons of coal is used by power plants and since power plants mostly use steam coal, this results in power plants consuming about 85% of global steam coal production.

Hard coal, here mostly coking coal, is also a crucial ingredient for two-thirds of global steel production. Other industries, such as alumina, paper, and chemical industries, also rely heavily on coal. Coal liquefaction has been perfected by Sasol, the South African company that uses the famous German Fischer–Tropsch technology to generate fuel from coal (for more details on coal-to-liquid processes, see Section 4.7). Sasol already consumes one-quarter of South Africa's coal production for the purpose of CtL (see Section 3.4.4).

Since internationally traded steam coal is mainly used for power generation, the following sections will focus on power generation and its implications for the global seaborne steam coal trade. I will analyze environmental issues associated with the use of coal, compare coal with other sources of energy for electricity generation, and begin discussing the future of coal use. Where appropriate I will draw from and refer to existing scientific knowledge. The purpose of this section is to present an overview of power generation and to draw conclusions for the coal industry that are relevant for the economical analysis of the coal trade market.

4.2 Steam Coal and Its Role in Power Generation

As discussed earlier, coal is used to generate over 40% of the world's electricity (see Fig. 3.1). As such, coal is the single most important source of energy for all of the world's electricity needs, outstripping gas (20%), nuclear (15%), and oil

(7%). Based on market research, I estimate that over 90% of global seaborne steam coal is used in power generation. This is confirmed by the German Coal Importers Association (VDKI 2006, p. 8). Thus, power generation drives the demand for seaborne traded steam coal.

Power generation requires availability and reliability of the energy resource. In addition to securing the raw material, power plants need to ensure they can get the produced electricity to the consumer. Coal, in many aspects, is the most reliable and widely available energy source for power generation. Major blackouts such as those that occurred in the last decade (for instance, in Brazil in 2001, in California in 2000–2001, in New Zealand in 2001 and 2003, in the Northeastern United States in 2003, in Italy in 2003, in China in 2008, and in South Africa in 2008) can be avoided in the future if steam coal is used more wisely, often in connection with renewable sources of energy (World Coal Institute-Secure Energy 2005). Many countries that currently use little coal for power generation, such as New Zealand and Brazil, are advised to increase their coal share in order to reach higher levels of system reliability.

The following sections will discuss steam coal demand by region and in the most important steam coal-importing countries.

4.2.1 Steam Coal Demand by Region

The world produced about 18,300 TWh of electricity in 2005. Australasia accounted for almost one-third of global power generation, while North America accounted for 28% and Europe plus Russia 24%. South America, Africa, and other countries accounted for the remaining 16%. Figure 4.1 summarizes the global situation in more detail.

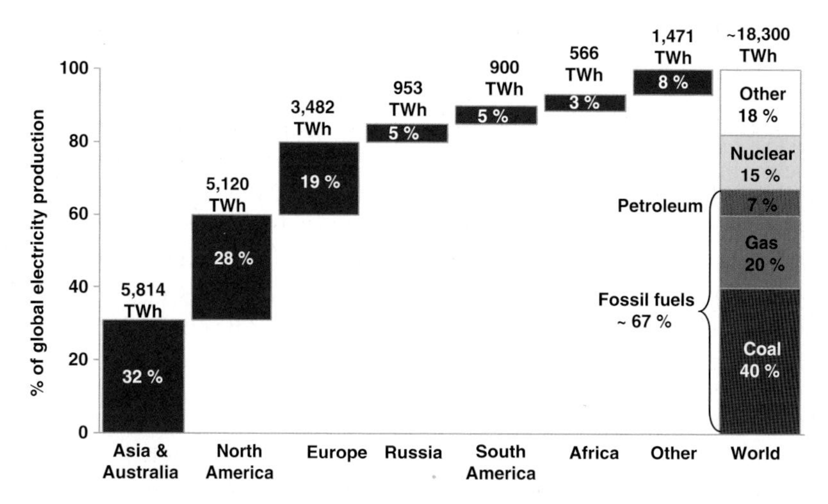

Fig. 4.1 World's electricity energy share by region in 2005 (Source: IEA-Statistics 2005; Author's analysis)

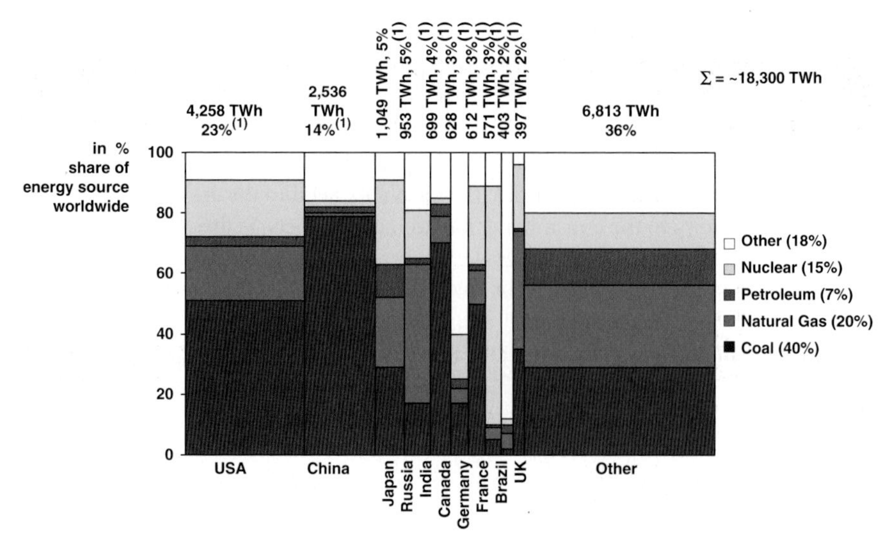

Fig. 4.2 Global top 10 electricity producers in 2005 ([1] Share of total electricity production in various countries. Source: IEA-Statistics 2005; Endres 2008; Author's analysis)

More interestingly, Fig. 4.2 shows the world's top 10 electricity producers and their respective power generation mix. The top 10 countries account for two-thirds of global power generation. China, India, the United States, and Germany generate more than half of their electricity using coal; in fact, China depends on coal for up to 80% of its power. The graph in Fig. 4.1 gives a good visual overview of the power generation mix. Canada uses a lot of hydro while France is the country with the highest nuclear power generation share of almost 80%. Japan, the world's third-largest power producer with a 27% coal power generation share, is also the single largest steam coal importer in the world because it has no significant coal reserves itself.

Appendix B examines power generation in all regions of the world and details the power mix of 26 of the most important countries in the world.

Looking to the future one can visualize, based on Fig. 4.2, that China's and India's columns will grow much wider and, thus, the world's use of coal will increase. Both countries rely more heavily than average on coal for their power generation mix; thus, their use of coal will increase more than that of any other fossil or even nuclear energy resource. This is confirmed by IEA Energy Outlook 2007, summarized in Table 4.2, which forecasts in its reference scenario that world primary energy based on coal will grow faster (2.2% per annum) between 2005 and 2030 than all other sources of energy except 'other renewables.' Coal's global share of primary energy is therefore expected to increase from the current 25 to 28%, reflecting the title of this study, '*The Renaissance of Steam Coal.*'

Not all countries have energy resources available domestically. Therefore, trade in natural energy resources is necessary and will continue to increase. Japan, South

Table 4.2 IEA forecast of global primary energy demand, 1980–2030

Mln toe	1980		2005		2030		CAGR 2005–2030 (%)
Coal	1,786	25%	2,892	25%	4,994	28%	2.2
Oil	3,106	43%	4,000	35%	5,585	32%	1.3
Gas	1,237	17%	2,354	21%	3,948	22%	2.1
Nuclear	186	3%	721	6%	854	5%	0.7
Hydro	147	2%	251	2%	416	2%	2.0
Biomass or waste	753	10%	1,149	10%	1,615	9%	1.4
Other renewables	12	0%	61	1%	308	2%	6.7
Total	7,228	100%	11,429	100%	17,721	100%	1.8

Sources: IEA Energy Outlook (2007, p. 74); Author's research and analysis

Korea, and Taiwan are today the largest importers of steam coal, accounting cumulatively for 38% of global imports (VDKI 2006). The following section will briefly examine the most important steam coal-importing countries as they form a key component in worldwide steam coal demand and therefore greatly influence the economics of the coal trade.

4.2.2 The World's Most Important Coal-Importing Countries

Of the 595 million tons of steam coal that was traded by sea in 2006 (see also Fig. 2.6), 114 million tons, or 19%, was imported by Japan; 61 million tons, or 10%, by South Korea; and 54 million tons, or 9%, by Taiwan. By comparison, Europe in total imported 191 million tons, or 32%, of the global seaborne steam coal trade in 2006. In Europe, the largest importers are the United Kingdom with 38 million tons, or 6% (Global Insight-World 2007, Section 2, p. 18), and Germany with 32 million tons, or 5%, of the global seaborne steam coal trade. Table 4.3 summarizes the import

Table 4.3 Coal import dependency of key coal-importing countries, 2006

Country	Percentage of global seaborne steam coal trade (%)	Import dependency rate (%)	Percentage of imported steam coal in electricity production (%)
Japan	19	99	24
South Korea	10	95	33
Taiwan	9	100	53
United Kingdom	6	63	21
Germany	5	69	14
United States	<1	2	1
China	<1	11	9
Spain	<1	71	17
Italy	<1	99	14

Sources: Haftendorn and Holz (2008, p. 2); Author's research and analysis

dependency of key coal-importing countries. The three biggest importers – Japan, South Korea, and Taiwan – rely entirely on imports while the two largest European importers rely on imported coal to supply two-thirds of their demand.

With the three Asian countries Japan, South Korea, and Taiwan importing 229 million tons of steam coal (38% of global trade), thus more than all of Europe, it is worth looking at these countries in more detail. I will also examine the United Kingdom and Germany. These top five countries account for almost three-quarters of the global steam coal trade. In 2006, the world's most populous nations China and India were not yet major steam coal importers. However, as this is likely to change in the near future, I will also briefly discuss these two countries.

4.2.2.1 Japan

Japan is often credited with having engendered the 'Asian industrial revolution' in the second half of the last century. Because Japan does not possess any significant natural resources itself, it was able to accomplish this through building up long-lasting and successful trading relationships with most of the world's major natural-resource-exporting countries. This is also true for the coal trade. Today, Japan is the single largest importer of steam coal. With 114 million tons of steam coal imports in 2006, Japan alone accounts for almost one-fifth of the global seaborne steam coal trade. Its share of the coking coal trade is even more substantial. Japan, which is the world's second largest steel producer after China, imported 63 million tons of coking coal, or one-third of all sea-traded coking coal (IEA-Statistics 2005; Global Insight-Russia 2007).

Japan's electricity generation is actually only based up to 29% on coal, while 28% is based on nuclear and 23% on gas (IEA-Statistics 2005). Global Insight-Russia (2007, Section 2, 10) confirms that there is continued support in Japan for more coal-fired power generation as a result of the country's ongoing nuclear difficulties, exemplified by public opposition, low utilization rates, and the shutdown of the Kashiwazaki Kariwa plant. It is expected that Japan's coal use will continue to increase in relative importance and peak in the year 2025. Thus, this country will continue to grow from a large basis and fuel Pacific steam coal demand. Japan, of course, signed the Kyoto Protocol.

About 50% of Japan's steam coal imports come from Australia and the rest mainly from Indonesia, Russia, and China. Since it is expected that Chinese coal exports will decrease over time (see Section 3.4.6 for detailed discussion of this issue), any supplier that can increase exports in the future will benefit from Japan's growth. Japan is one of the highest-paying countries, but also one of the choosiest when it comes to coal qualities.

4.2.2.2 South Korea

The second largest steam coal importer, South Korea, trails Japan by almost 50%. Nevertheless, South Korea's demand was a key factor influencing the price spikes in 2007 and 2008. South Korea imported 61 million tons of steam coal in 2006 and depends on coal for 38% of its electricity generation. This compares with 37% from nuclear and 16% from gas power generation in 2005 (IEA-Statistics 2005, Appendix 2). Up to 2010, Global Insight-Russia 2007 (Sections 2, 13) forecasts an increase in coal consumption translating into an import demand of up to 80 million tons. In the longer term, it is expected that the relative share of coal in South Korea's power generation mix will fall and imports will stagnate or decline slightly. South Korea did not sign the Kyoto Protocol.

The key suppliers to South Korea are, like in the case of Japan, Australia, Indonesia, Russia, and China. However, in 2007 the first imports of South African coal were noted. This means that Asian countries are now competing for coal that historically has been sold almost exclusively to Europe. Thus, even if European demand does not increase much, there are signs that supply will be under pressure in the medium to long term.

4.2.2.3 Taiwan

Taiwan imported 54 million tons of steam coal in 2006, almost as much as Korea. Coal accounts for 54% of the country's power mix compared to 18% for nuclear and 17% for gas. It is expected that Taiwan will continue to build and expand its coal-fired power plants and further increase steam coal imports, reaching almost 70 million tons by 2025 (Global Insight-Russia 2007, Sections 2, 14).

Taiwan holds a special position in the Pacific coal market. The country is logistically much better situated than Korea or Japan because of its proximity to Indonesia. However, it is unclear how long Indonesia can continue its export increase. For political reasons, Chinese coal exports to Taiwan have been unthinkable. Australian coal has also been a key source for Taiwan. Given its flexibility on quality, Taiwan, unlike Japan, can look to a wider range of import sources in the future.

4.2.2.4 Europe: United Kingdom and Germany

Germany and the United Kingdom are Europe's most important steam coal importers. With 70 million tons of imports in 2006, the two countries together accounted for 12% of global seaborne steam coal demand and 37% of total European seaborne imports (VDKI 2006). The two countries are at the core of the Atlantic market and their development will have a major impact on future trade flows and price developments, especially since the world's highest marginal cost producers, Russia and Poland, export mostly to these two countries.

United Kingdom

Coal accounts for 35% of the United Kingdom's power mix, compared with 39% for gas and 21% for nuclear. It is expected that coal will decrease in relative importance in the power mix. The government supports new nuclear power stations and overall electricity demand is not expected to increase much, if at all. The import growth in the United Kingdom, if at all, will come from decreased indigenous production;

this totaled 19 million tons in 2006 (VDKI 2006, p. 62). Domestic production is expected to fall to 15 million tons by 2020 while imports are expected to fall from the current 44 million tons to 35 million tons within the same period (Global Insight-Russia 2007, Sections 2, 18). Almost half of the United Kingdom's imports are sourced from Russia. It is expected that the projected import decline in the United Kingdom will have little negative effect on prices as Russia is likely to increase exports to the Far East and overall Russian exports are at risk of declining in the medium to long term (see also Section 3.4.5).

Germany

Germany still relies heavily on coal: About 51% of the German power mix is based on coal products (about half each on lignite and steam coal). Nuclear accounts for 27% and gas for 11%. Germany has invested heavily in wind power; coal power plants are already required to shut down in periods of strong wind, as government regulations give preference to nonfossil-based electricity irrespective of its generation costs.

Local production totaled about 21 million tons in 2006 and the government has decided to phase out all subsidies by 2018 at the latest. The German Coal Importers Association (VDKI) and various other sources predict that Germany's steam coal imports will increase substantially, reaching 40+ million tons by 2015. Germany is one of the few European countries that strongly supports the building of new coal-based power generation capacity. Coal-fired power plants with a planned capacity of 15.7 GW hold the largest share of the 53 known power plant projects totaling 31.4 GW (Argus Coal Daily and VDEW, 19 October 2006, p. 5). In parallel, nuclear power generation faces strong opposition from the left-wing and Green parties, which potentially opens up additional opportunities for gas- and coal-fired power plants, even if that is actually not the goal of such political pressure. However, in the medium term, especially after the 2009 election results, we can conservatively expect Germany to retain the nuclear share of its power mix.

4.2.2.5 China and India

In the previous sections we discussed the world's largest steam coal importers. In total, the five largest importers (Japan, South Korea, Taiwan, Germany, and the United Kingdom) accounted for 71% of seaborne steam coal imports in 2006. The two most populous countries in the world, China and India, which together account for almost half of the world's population and which show continued strong growth, are not yet a significant demand component. However, China produces 79% of its electricity with coal and India 70% (IEA-Statistics 2005). Both countries use their domestic coal production to cover most of this demand. It is fair to assume that China and India will play an increasingly important role in the future demand for seaborne steam coal.

India's imports of steam coal alone are projected to increase from around 20 million tons per annum in 2006 to over 100 million tons per annum by 2020. India has the world's fourth largest coal reserves (BGR 2006), but recent acquisitions of global mining assets by Indian companies, such as Tata, in Indonesia, are a sign that India could become one of the world's largest coal-importing countries in the medium term. India's reserves will dwindle significantly and the Indian Ministry of Coal expects that reserves will be depleted within 20–30 years (PESD 2009, P. R Mandal). In 2007, India became the single largest importer of South African coal, a development that would have been unheard of only a year earlier. We already discussed China's role as a swing supplier in Section 3.4.6. It should suffice to mention here that China will have a huge impact on the global market if it turns to significant steam coal imports as most in the coal trading industry expect.

It can be concluded from Section 4.2 that overall demand for seaborne traded steam coal will increase in the future. Growth will be driven by Asian countries, but also Europe's demand is expected to remain at least constant. It is expected that the Pacific coal market will further outpace the Atlantic market. Suppliers to the Atlantic market such as South Africa and Russia are already adjusting. Latin American and North American suppliers will also have to start looking at their long-term strategies and discuss possibilities for exporting not only toward the east but also toward the west. In the case of North America, Peabody and Arch Coal are already evaluating options for coal exports from the Powder River Basin (PESD 2009).

In the following sections, I will look at some more technical aspects of coal-fired power generation and its impact on the trade market.

4.3 Introduction to Power Markets

The power market is relatively complex due to certain product characteristics. These include the following: (1) without energy there can be no economic activity on this planet (wars have even been fought over access to energy); (2) fossil energy sources are geographically concentrated and those geographical locations rarely match the locations for energy demand; (3) transforming energy from one form to another often causes environmental pollution; (4) mining and power generation can be extremely dangerous (e.g., risks associated with nuclear power plants); (5) electricity generation requires large investments with a very slow return; (6) in many, if not most, countries, energy resources, power generation, and the grid are controlled by the national government; and, last but not least, (7) – often a result of all of the above – energy and power markets are far from economically efficient; perfect competition does not exist.

The product of the power market, electricity, also has certain characteristics that further complicate the market (see Erdmann and Zweifel 2008, pp. 7 and 293):

- Electricity is extremely difficult to store, at least for industrial applications.
- Electricity can be transported cheaply across long distances, but it is bound to a grid or other system for transportation.
- Electricity carries energy with maximum exergy; thus, most electrical applications have a very high efficiency.

- The laws of thermodynamics do not limit density of energy; therefore, very high temperatures can be reached.
- Electricity has practically no mass or volume; thus it is well suited for communication purposes.

The following sections give a brief overview of power supply and demand, matching power pricing and the competitive conditions under which at least European utilities operate. I focus on these because they are relevant for the steam coal market, the focus of this study.

4.3.1 Matching Power Supply and Demand

As a result of many of the complexities of the power market mentioned earlier, it is a challenge to match electricity demand with electricity supply.

Figure 4.3 illustrates exemplary electricity demand for a typical summer and winter day in Germany. The industry has optimized the utilization of various electricity generation methods to best match the demands of base, intermediate, and peak loads. The definition is arbitrary but makes sense in terms of operating hours for a power plant serving the various load types. Table 4.4 summarizes the hours that various power generation facilities typically run every year and when these facilities are usually used. Thus, utilities choose to invest in power plant projects depending on the predicted capacity for the various load ranges they would like to fill. Given the long lead time of power plant investments, this often proves very difficult. In many countries, the operating hours of hard coal power plants, however, may vary significantly from the hours in the German market. For example, coal is used for base load in South Africa, Poland, Australia, China, India, and many other countries around the world.

The usage pattern of the various power generation facilities also depends on the marginal cost of operation. For instance, wind, nuclear, and hydro have very low

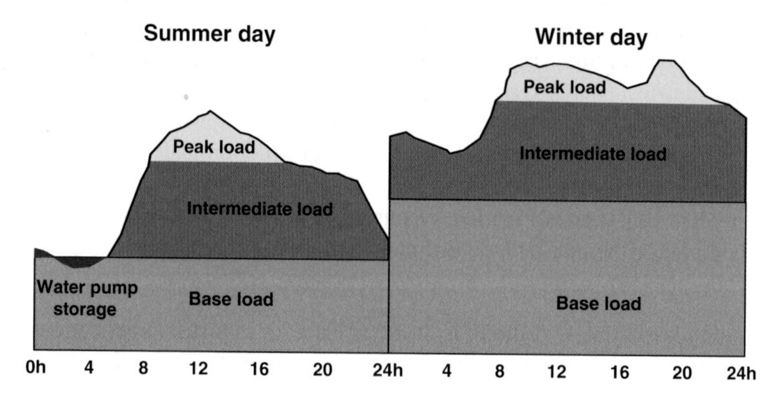

Fig. 4.3 Exemplary daily load for a summer and winter day (Source: VDEW 2007)

Table 4.4 Typical usage pattern of power generation facilities in Germany

Type of generation facility	Typical operating hours per annum	Typical load usage
Wind mill	Depends on wind	Base load because of law
Nuclear power plant	7,500 h	Base load
Lignite power plant	7,000–7,500 h	Base load
Hydro power plant	4,900 h	Base load, not more operational hours because of water levels
Hard coal power plant	4,800 h	Intermediate load
Gas power plant	2,400 h	Peak load
Pump storage hydro power plant	1,000–1,500 h	Peak load

Note: 1 year has 24 h × 365 days = 8,760 h
Sources: VDEW (2007); Schwarz (2006); Author's market interviews

marginal costs. Hard coal tends to have higher marginal costs given the cost of the raw material, while oil and gas power plants tend to have the highest marginal costs. The next determinant for the usage pattern is the time it takes to fire up a power plant. Peak load facilities require a swift response to increased or decreased power demand. As an indication we can state that it takes about 2–4 h to start up a nuclear power plant, about 30 min for a coal-fired power plant, and 5 min or less for a gas facility. However, some modern coal-fired plants can switch load levels relatively quickly, also within 5–10 min.

4.3.2 Power Pricing and Coal

It is difficult to discuss this complex subject in brief. I will start with the results of an analysis performed by Schwarz (2006) that found no long-term and consistent correlation between EEX (European Energy Exchange) power prices and API2 coal prices. Schwarz analyzed the period 2002–2006 in her study, which was performed with the help of E.ON, one of Europe's leading utility companies. This is confirmed by Erdmann and Zweifel (2008, p. 264), who analyzed the period 2000–2006.

In terms of pricing, in order to hedge risks, utilities sell a large part of their generation capacity through long-term contracts with municipalities and large industrial organizations. The part that is not sold long term is sold on the spot market either via

- exchanges, such as the EEX in Germany; or
- over the counter (OTC), through bilateral contracts between any two parties; or
- through brokerage platforms.

On the spot market, market participants trade electricity a day ahead either in individual hours or in blocks of hours. Since the marginal cost curve for electricity generation is very convex, even small changes in demand can result in very large

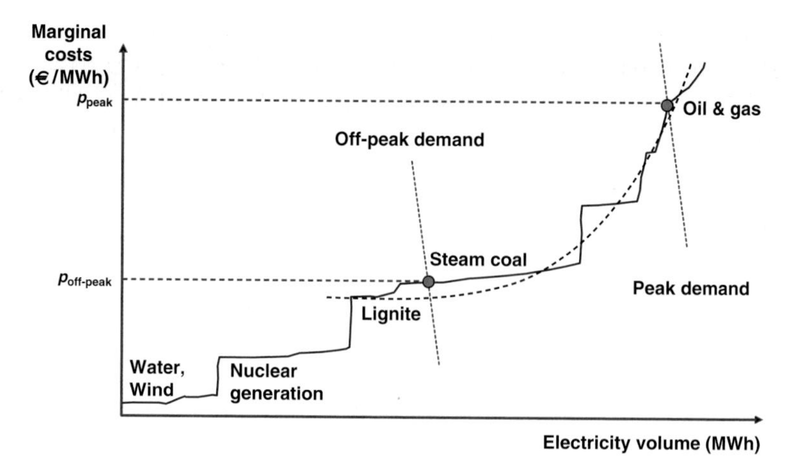

Fig. 4.4 Dispatching and exemplary pricing on spot power markets (Source: Erdmann and Zweifel 2008, p. 304)

price fluctuations, especially when the demand changes occur during peak hours. See Fig. 4.4 for an illustration of a typical day-ahead spot market indicating pricing during off-peak and peak hours.

The difference between the steam coal price and the base load power price is called 'dark spread' (DS). To calculate the dark spread, market participants assume average coal power plant efficiencies of 38%. The clean dark spread (CDS) takes into consideration the price of CO_2 emissions assuming the power plant requires 0.89 tons of emission rights for each MWh (see Erdmann and Zweifel 2008, p. 266 and author's calculations).

Erdmann and Zweifel (2008, p. 266) have identified statistical significance in the correlation between the clean dark spread and the power price. They analyzed European coal prices and German power prices between September 2000 and November 2006.

Power generators can only earn money or cover their fixed costs if the clean dark spread is positive. 'Free' emission rights distributed by the government to utilities have no relevance in determining the price for power (they do, of course, have major relevance for the profits of the receiving utility). Power prices depend on the price for emissions but do not depend on the number of 'free' emission rights distributed. This is because the price depends only on the cost of one marginal unit of power generated, thus including costs for a marginal ton of emissions.

Serious steam coal market participants look at the dark spread and spark spread on a regular basis. Clean dark spreads that are positive for a long period of time give an indication that either electricity prices are likely to fall or there is room for coal prices to increase. The same is true the other way around. For example, on June 5, 2008, European coal prices were at record price levels, hovering above US $170.00/ton basis 6,000 kcal/kg CIF ARA (Argus Coal Daily, June 6, 2008). On that day, however, German clean dark spreads were negative at –€10.27/MWh for base load and positive €22.03/MWh for peak load. UK clean dark spreads, on the other hand, reached almost record levels at 9.10 GBP/MWh for base load and 29.80 GBP/MWh for peak load. Thus, despite fuel prices having almost tripled in one year, clean dark spreads were still healthy in the United Kingdom but did not look so good in Germany. The message to the coal market was that electricity prices would not limit further coal price increases in the United Kingdom, but that Germany could not sustain current fuel price levels unless electricity prices increased further.

4.3.3 Competitive Conditions of Power Generation

The power markets have changed significantly in the past few years, especially in the United States, Europe, and the Far East. Traditionally, the power industry is a monopolized, state-controlled business. There are several arguments in favor of state control of power generation (here I have only listed a selection, please also see Ritschel and Schiffer (2007, p. 41) and Erdmann and Zweifel (2008, p. 322)):

- In order to supply stable and reliable electricity to every household and industry operation, the power system needs to be designed with significant overcapacity to handle peak times and a little extra for 'extraordinary' times. Private companies, whose interest is in maximizing profits, have no economic incentive to hold this overcapacity.
- The power system is one of the most important building blocks for a healthy economy. A weak power system weakens the country, thereby draining the nation's wealth.
- Liberalized power generation requires liberalized power grid operation. However, power grid operation is a so-called natural monopoly, which means that one company will always be more efficient than two companies handling the same volume. Thus, well-thought-out regulation is needed to avoid owners taking advantage of their monopoly power on power grids.
- Power plant investments have a very long time horizon (usually a minimum of 20 years), which few companies are able to sustain.

Nevertheless, liberalization, deregulation, and unbundling have led to a reduction in monopoly structures in the West and in some parts of Asia. Today, in most European countries we find an oligopolistic competitive structure in power generation (seeTable 4.5). This, one may argue, is a step forward from a monopoly toward more efficient markets. Pricing in this market therefore involves a classic application of the Cournot game or Cournot/Bertrand competition. I discuss these game theory models in more detail in Chapter 6.

Table 4.5 Power generation market concentration in Europe

Country	Largest generator (% of country's capacity)	Top three generators (% of country's capacity)
United Kingdom	20	40
Germany	30	70
France	85	95
Italy	55	75

Note: Data as of 2003.
Sources: European Commission (2005)

4.4 Coal-Fired Power Plant Technologies

Coal power plant technology is considered to be one of the most stable technologies for generating electricity. This in fact is one of the many advantages of coal-based electricity over other energy sources. For a detailed discussion of power plant technologies I can refer the reader to the IEA-Manual (2006), one of many publications that explain this subject. Instead of discussing all technical details here, I will focus on a brief selection of economic and environmental issues with relevance for the international seaborne steam coal market. It is interesting, but in terms of efficiency unfortunate, that there are scarcely two coal power plants in the world of the same design. This means that almost every power plant is different and also works differently, resulting in varying efficiencies among power plants when burning various qualities of coal.

4.4.1 Power Plants and Coal Use

In a coal-fired power plant, the chemical energy stored in the coal is first converted into thermal energy in the boiler, then to mechanical energy in the turbine, and finally to electrical energy using the generators. Figure 4.5 depicts the typical process of a coal-fired power plant.

4.4.1.1 Coal Preparation

The coal first needs to be prepared for the boiler. This depends on whether the boiler is fed the coal using pulverized injection or by the traditional method using screens. Thus, the size and hardness of the coal (see HGI above) are important at this stage. Modern power plants use mills that sustain more wear if the coal is harder. The feeding system may also have trouble with large-sized coal. Therefore, typical steam coal is supplied in the size 0–50 mm; however, if it wasn't for the problem of dust, most power plants would happily take finer coal (i.e., 0–6 mm) to save strain on their mills.

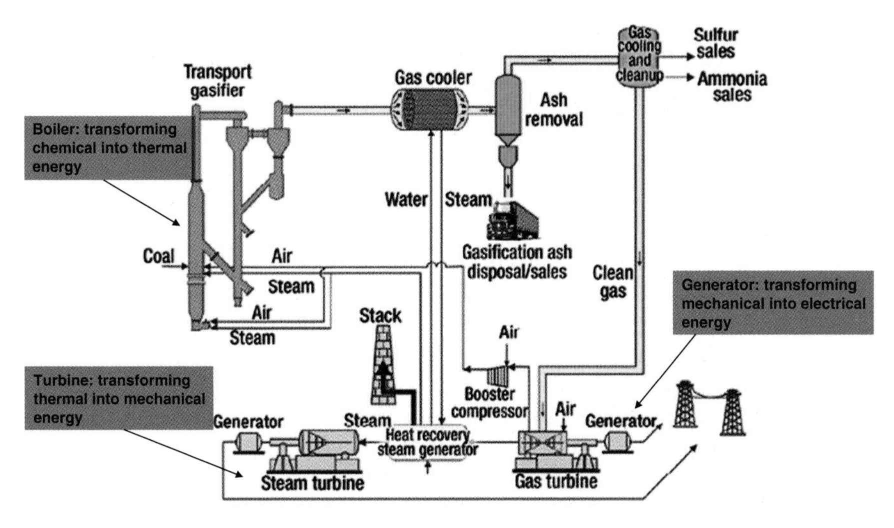

Fig. 4.5 Typical process in a coal-fired power plant (Source: http://www.enr.com)

4.4.1.2 Coal Burning

The coal is then burnt in the boiler, typically at temperatures between 900 and 1,200°C. Here the chemical energy contained in the coal is transformed into thermal energy contained in steam. The various characteristics of coal, including but not limited to the calorific value, volatile matter, moisture, and ash fusion temperatures, were already discussed in Section 3.3. It should suffice to say that this is the stage where the type of coal can make the greatest difference in terms of the resulting power plant efficiency. Technical systems need to be adjusted to specific types of coal in order to produce optimized power output. Table 4.6 summarizes the typical

Table 4.6 Costs for a reference coal-fired power plant in Germany

Cost category	Amount (2002/2003)
Installed capacity	600 MW
Investments	€798/kW (gross)
Captive requirements	7.4% of gross capacity
Captive requirements	44.4 MW
Maintenance	1.5%/year
Personnel	70 people
Fuel costs without CO_2	€55/t SKE
Fuel costs	2.1 cents/kWh
Total marginal cost	~4 cents/kWh

Source: *Studie Referenzkraftwerk Nordrhein-Westfalen* [Study of Reference Power Plant in Nordrhein-Westfalen], VGB PowerTech e.V., Essen, http://www.vgb.org; VGB 2003

costs of a reference coal-fired power plant in Germany. However, fuel costs and total marginal cost per kWh can vary significantly depending on the type of coal used.

The type of coal has no impact on the next process stages, consisting of transforming thermal energy into mechanical energy in the steam turbines and transforming mechanical energy into electrical energy in the generators.

The chemical characteristics of the coal does influence the efficiency and capacity of auxiliary equipment such as the wet flue gas desulfurization (WFGD) equipment, as discussed previously. It is recommended that coal market participants visit a power plant at least once to understand some of the key technical aspects of burning coal.

I personally have witnessed cases where a higher-quality coal that may cost US $2.00/ton basis 6,000 kcal/kg NAR more than another coal of lower quality saved more than US $8.00 a ton – calculated back to a per-ton price – during burning. Therefore, it is important to remember that (1) not every power plant can burn every type of coal and (2) the price of the coal is only one component in the marginal cost of a power plant; efficiency gains resulting from burning the right coal quality are often more important.

While power plants need to be as flexible as possible in terms of accepting various coal qualities for their burners, the purchasing department should look for the right coal quality for their specific boiler on the market at all times. Frequently there is a misalignment of incentives, resulting in quarrels between technical teams and commercial teams at power plants. Despite the fact that I come from a commercial background myself, I would argue that commercial procurement departments can go a long way toward making utilities' operations more efficient, thus saving a lot of money. Power generators have to rethink their incentive systems; rather than measuring only the cost of coal per ton, they should also start looking at how the procured coal behaves in the power plant.

4.4.2 The Future of Power Plant Technology: Increased Efficiency, Reduced CO_2, and CCS

Engineers are currently focusing on two principal means of optimizing coal-fired power plant technologies: (1) by increasing efficiency and (2) by reducing CO_2 emissions. However, these two aims compete to an extent; reducing CO_2 emissions always reduces efficiency, as energy is required to capture and process CO_2 for storage.

Figure 4.6 indicates current average efficiencies and the resulting CO_2 emissions. Figure 4.2 shows the amount of coal-fired electricity production in China. It can now be derived that the driver for CO_2 reduction is not the western world, but China and other developing countries. If China reached the same technological standard as Germany, then CO_2 emissions in China from coal-fired power generation could almost be halved. A total of 1.1 billion tons of CO_2, or over 4% of the 26 billion tons

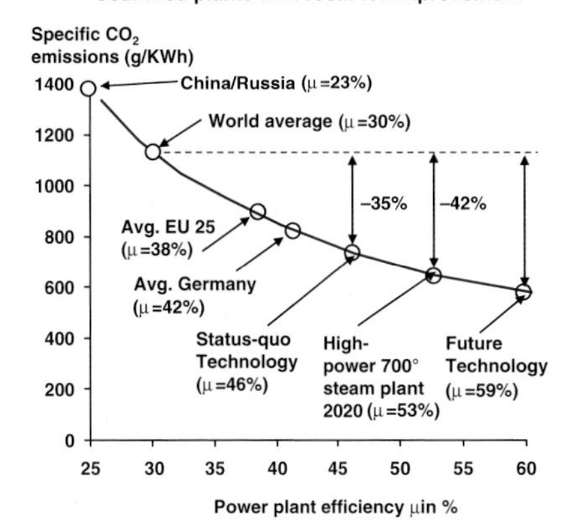

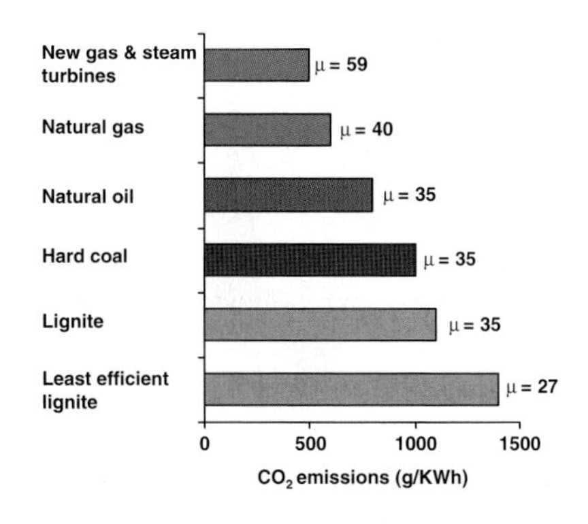

Fig. 4.6 Efficiency and emissions of coal power plants (Source: Deutsche Bank 2007; UBS-Resources 2008; Author's research and analysis)

of CO_2, emitted globally through human activity could be saved by this measure alone.

Future technologies are expected to increase conversion efficiencies to 60% or more. However, this has little impact on the present challenge of capturing and storing carbon dioxide. For a more detailed discussion of carbon capture and storage (CCS) please refer to a series of research papers from Stanford University (Rai et al. 2008). Engineers are currently working on three principal technologies for CCS.

4.4.2.1 Flue-Gas Scrubbing in Conventional Power Plants

The advantage of this process is that it can be used with any existing conventional power plant. The major disadvantage is that the efficiency is reduced from about 42% (in Germany) to 28%, a large sacrifice. The technology works by separating the CO_2 from the dedusted and desulfurized flue gas in an additional scrubbing stage under atmospheric pressure.

4.4.2.2 Oxyfuel Process

In this process the combustion of the fossil fuel occurs not with air but with recirculated CO_2 and oxygen. Since the flue gas now consists largely of CO_2, it can be captured without any additional scrubbing stage; thus, power plants using this process can currently reach efficiencies of up to 37%.

4.4.2.3 Integrated Gasification Combined Cycle (IGCC) Process

This process is currently the most promising. The CO_2 is actually captured before combusting the fuel. The fuel is gasified, the CO_2 is separated, and then the gasified fuel is combusted in a gas turbine. The IGCC process currently reaches efficiencies

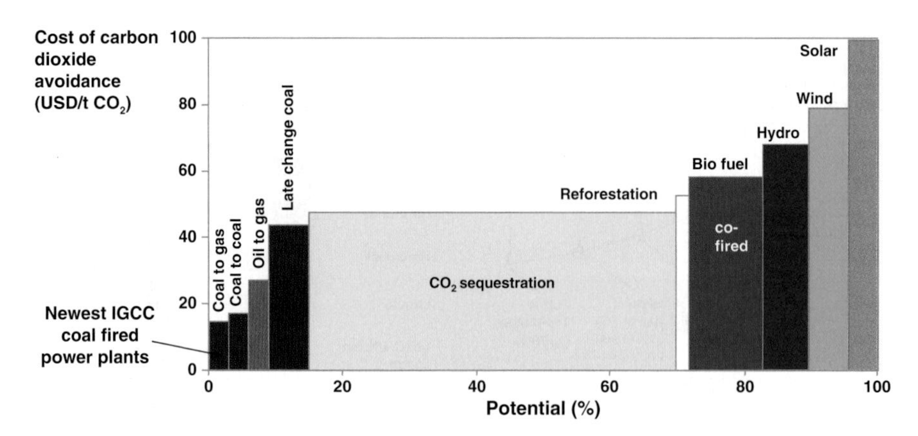

Fig. 4.7 Exemplary specific CO_2 avoidance costs (Source: BCG 2004)

of up to 40%, thus almost reaching the current standard of power plant efficiency. This process is also the least expensive; nevertheless, Ritschel and Schiffer (2007) estimate that the investment required for an IGCC power plant is up to 80% higher than that for a conventional power plant.

Given the above technologies, coal-based power generation offers relatively low CO_2 avoidance costs. The Boston Consulting Group (BCG) has summarized in Fig. 4.7 a vast array of CO_2 avoidance technologies. While scientists and the power industry argue that it is extremely difficult if not impossible to accurately quantify CO_2 avoidance costs, it does appear likely that inexpensive methods (or simply efficiency gains) will come from new IGCC coal-fired power plants. This single fact supports the theory that steam coal will remain the most important source of power for the coming decades. Most opponents of coal-fired power are unaware that much of the CO_2 problem can be dealt with in an efficient way. As this awareness grows I also forecast a growing coal trade market.

4.5 Environmental Issues Associated with Coal Use

Coal-fired power generation faces most criticism for its high CO_2 emissions, and rightly so. Currently 74% of all CO_2 emissions from power generation are caused by the combustion of coal, while only 17% are caused by gas. Therefore, coal accounts for about 40% of the 26 billion tons of anthropogenic CO_2 released into the atmosphere (IEA-CO_2 2006) (see Fig. 4.8). The World Wide Fund (WWF) for Nature published a report entitled ‘Dirty Thirty’ listing the most CO_2-polluting power plants in Europe. The results were not surprising: only one of the 30 co-fires gas with coal, two co-fire oil with coal, and all others burn only coal (WWF 2007). The worst three plants are lignite-fired power plants in Greece and Germany. For more details on global CO_2 emissions, please refer to Appendix C.

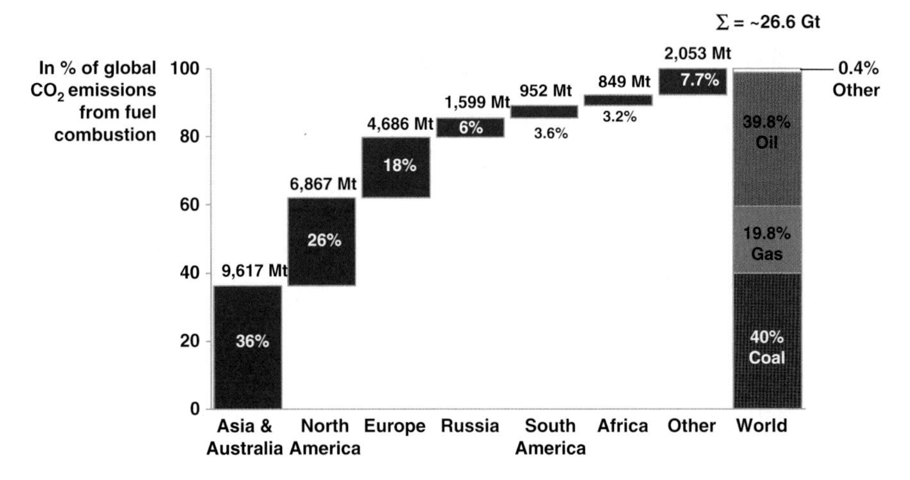

Fig. 4.8 Global CO_2 emissions by source (Source: IEA-CO_2 2006; Author's research and analysis)

However, CO_2 emissions are not the only negative effect that coal-fired power generation has on the environment. Pollutants such as nitrogen, sulfur, and the trace element mercury have also been a problem for the industry. Environmental issues associated with the production of coal have already been discussed in Section 3.6.5. In addition, the transportation of coal causes further emissions, which are statistically attributed to oil. But first, I will discuss the most relevant and talked-about pollutant stemming from coal, CO_2.

4.5.1 How Much CO_2 Stems from Coal?

Before discussing this subject, it is worth pointing out that CO_2 is nontoxic; indeed, without it no life would be possible on Earth. However, too much CO_2 is a major cause of the well-studied 'greenhouse effect,' which in turn leads to problems similar to those experienced in previous climatological ages. The major difference is that this time the change is caused by human behavior. Human development is a part of evolution, and evolution has caused the planet's climate to change more than once in the five billion years of the Earth's existence. Evolution brought the human race into existence, and for the first time in the planet's history a species – the human race – is capable of directly influencing nature. However, this is a philosophical discussion. There is no question that engineers should do everything to avoid the emission of CO_2 from coal-fired power plants, but we cannot expect them to be capable of reducing the emissions to zero.

The greenhouse effect is causing the Earth's average temperature to rise. Scientists estimate that the current temperature increase is the fastest in 20,000 years. The highest known Earth temperature (determined based on analysis of marine sediments) was reached about 55 million years ago. The planet's most recent 'warm period' ended about 115,000 years ago. Cold and warm periods have been alternating for about 2.7 million years. Currently, we are in a warm period that began about 11,500 years ago. Humans, however, are contributing to this warm period in the form of the greenhouse effect caused among other things, by, CO_2.

How much CO_2 is actually released by burning coal? Burning 100 tons of standard South African steam coal with 6,000 kcal/kg NAR calorific value, 25% volatiles AR, 11% ash AR, 0,7% sulfur AR, and 65% carbon content AR will result in the following by-products:

- 238 tons of CO_2;
- 14.4 tons of ash;
- 1.3 tons of SO_x; and
- dust and some NO_x (depending on the nitrogen content).

Assuming power plant efficiency of $\omega = 0.38$, 1 MWh of coal power generation therefore emits 0.89 tons of CO_2:

$$\begin{aligned} &C_{\text{Mass}=12.011} + O_{2\,\text{Mass}=15.999} = CO_{2\,\text{Mass}=44} \\ &\Rightarrow 1\,\text{t C} = 3.67\,\text{t } CO_2 \\ &\Rightarrow 1\,\text{t coal} = 0.65\,\text{t C} = 2.36\,\text{t } CO_2 \end{aligned} \tag{4.1}$$

$$\begin{aligned} &1\,\text{t Coal at } \omega = 0.38 \Rightarrow 1{,}000\left(6{,}000\tfrac{\text{kcal}}{\text{kg}}\right)0.38 = 2{,}280\,\text{kcal} \\ &860\,\text{kcal} = 1\,\text{KWh} \Leftrightarrow 1{,}000{,}000\,\text{kcal} = 1{,}162\,\text{MWh} \\ &\Rightarrow 1\,\text{t Coal}_{\omega=0.38} \approx 2.65\,\text{MWh} \\ &\Rightarrow 1\,\text{MWh}_{\text{Coal}} \approx 0.89\,\text{t } CO_2 \end{aligned} \tag{4.2}$$

The best way to reduce CO_2 is to increase the efficiency of coal-burning power plants. Each 1% increase in average global coal-fired power plant efficiency would result in approximately 1% less carbon dioxide emitted. Considering that coal burning for electricity production resulted in 7.1 billion tons of CO_2 emissions in 2004, a 1% global efficiency increase would reduce CO_2 emissions by 71 million tons or 0.3% of all CO_2 currently emitted by human activity. (Note: in fact, the emissions reduction could be even greater as average global power plant efficiency reaches 30%; thus, it would depend on which base the efficiency increase starts from.)

4.5.2 Nitrogen and Sulfur

Nitrogen oxides contribute to global warming by helping to destroy the ozone layer. While the concentration of nitrogen oxides is much lower than that of carbon oxides, the global warming potential of nitrogen oxides is considered 300 times larger than that of carbon oxides. Higher than normal nitrogen levels have other adverse effects on human, animal, and plants.

Robert Davidson (1994) points out that it is difficult at present to predict the emissions of nitrogen oxides from coal combustion. It makes sense that the amount of nitrogen in coal, and the way in which it is bound into the coal structure, affects the amount and distribution of nitrogen oxide emissions. The relationship between nitrogen in coal and emissions of nitrogen oxides is an area in which further research is needed, especially if more effective in-furnace abatement techniques are to be developed. Some countries, especially the United States, already have strict NO_x limits that influence procurement of coal by utilities. Other countries (e.g., in parts of Europe and Asia) have no such limits yet in place.

Nitrogen oxides, like sulfur oxides, also contribute to acid rain. The health impacts of sulfur dioxide include asthma attacks, eye irritation, coughing, and chest pain. Sulfur contents are monitored carefully when procuring coal for combustion. In general, internationally traded coal has sulfur content below 1%. Higher-sulfur coal is burnt only after blending with lower-sulfur coal. Most power plants employ sophisticated desulfurization equipment to limit the amount of sulfur oxides released into the atmosphere. Modern power plants are designed in such a way that virtually pure steam is emitted.

I would like to end this chapter with the key messages from IEA's Coal Industry Advisory Board (CIAB) published in their paper 'Reducing Greenhouse Gas Emissions – The Potential of Coal' (IEA CIAB 2005). In the long term (>2020), CO_2 capture and storage offers the potential for near-zero CO_2 emissions from coal-based power plants. Delivery of this option requires coordinated research, development, and demonstration (RD&D) now. In the next decade, cost-effective reductions in CO_2 emissions can result from increased coal combustion efficiencies achieved through the more widespread use of state-of-the-art coal-based power plant technology. These strategies are complementary: Deployment of modern, efficient coal-fired electrical generation technologies in the short to medium term can enable carbon capture at less cost in the longer term, if those power units are designed to enable cost-effective carbon capture retrofitting when that technology becomes available for commercial application. Successful implementation demands that government and industry work together; commercial markets will not deliver without appropriate and stable policy frameworks.

4.6 Comparative Analysis of Coal Substitutes for Power Generation

In any market demand and prices for a product are also driven by the availability of substitutes and the costs of switching between substitutes. In this section I will examine other fuels and techniques for power generation, namely gas, nuclear, oil, and alternatives such as hydro, wind, biomass, and solar. Toward the end of this section I will compare the various fuels and techniques.

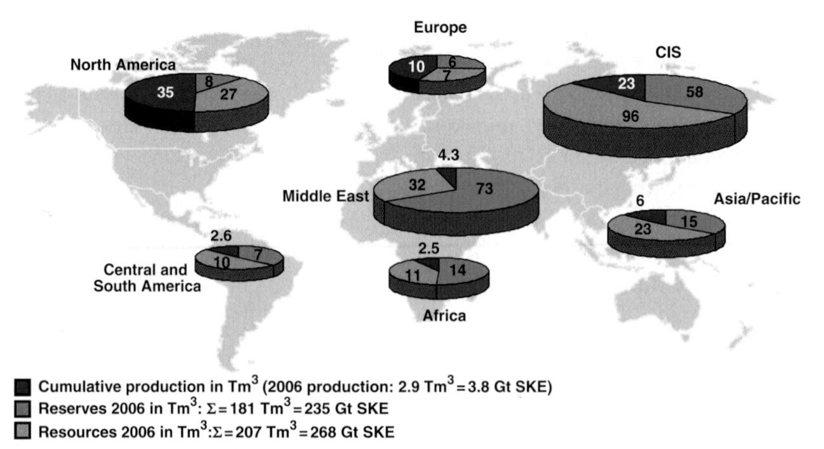

Fig. 4.9 Global natural gas production, reserves and resources (Source: BGR 2006; Author's research and analysis)

4.6.1 Natural Gas

Gas accounts for 20% of the world's electricity production and is the second most important fuel for electricity generation after coal, which accounts for 40% of global electricity production. Gas is also one of the most important sources of primary energy, accounting for 21% of primary energy in 2005 compared to 35% for oil and 25% for coal. For further details see Fig. 3.1 and Appendix B.

BGR (2006) estimates the remaining potential of natural gas at 388 Tm^3 (reserves at 181 Tm^3). Considering the 2006 production of 2.9 Tm^3 this translates into approximately 62 years reserve/production ratio, and 134 years when taking the remaining potential into account. However, gas reserves and resources are spread very unevenly across the globe. Russia and the Middle East have the largest available reserves. Over half of the world's natural gas reserves are situated in Russia, Iran, and Qatar, Fig. 4.9.

The reserves and resources of natural gas are large and we can expect that resources will continuously be reclassified in the years to come (however, please also refer to my notes on reserves and resources in Section 3.4.1). At the same time consumption will increase significantly. What makes gas more expensive than coal are the logistics involved in getting the gas to the power plant. The key advantage of combusting gas over combusting coal is that it emits approximately 45% less carbon dioxide than coal. Modern gas and steam cogeneration plants can reach efficiencies of 58% and more resulting in even further reduced carbon dioxide emissions and a more efficient use of this natural resource. However, the marginal cost of producing electricity in a gas-fired power plant is higher than in a coal-fired power plant.

There are two ways of transporting gas: (1) via pipelines, which is historically and currently the most common method, and (2) by liquefying gas and then transporting it as liquefied natural gas (LNG) before decompressing it again before use.

4.6.1.1 Gas Transport Via Pipelines

Gas pipelines have been in existence for a long time and have always played an important strategic role for governments and large companies. Investments associated with pipeline construction are very large; one can estimate approximately US $1–2 million/km, thus US $1–2 billion/1,000 km. When we consider that many pipelines exceed 5,000 km in length it becomes clear that a single company can rarely carry such large investment costs alone. For that reason, consortia are built so that ownership, investments, and risks are shared among a number of parties. Not surprisingly, then, there are a limited number of players involved in such large-scale projects. Once pipeline investments are shared, the operation of pipelines is also often shared, thus leaving room for a monopolized market driven by a cartel. NATO has already warned of such a Russian cartel many times, as has been reported by Financial Times Deutschland (2006).

Once the pipeline has been constructed, investment costs are sunk costs (see Erdmann and Zweifel 2008, 226ff). Pipelines are factor-specific, and thus can only be used for gas; they can only be moved to another location at great expense and depend on bilateral contracts with usually monopolized producers and consumers. This results in what is known as the *hold-up problem*, which means that the contract partners (producer and/or consumer) can take advantage of the strategically weak pipeline investor.

When the producer and importer cooperate, as is often the case in the real world, the result is a lower price for the consumer and higher profits for the producer and importer. Thus, cooperation or vertical integration in pipeline transportation is beneficial for everyone and in the interest of the economy as a whole, at least theoretically. In practice, for instance in a competitive market such as the European Union, it would be better for the consumer if the natural pipeline monopoly were regulated and if gas production were stimulated to be as competitive as possible.

Erdmann and Zweifel (2008) showed that Gazprom's behavior in diversifying its pipeline portfolio toward the west, thus controlling more than one pipeline route, leads to increased transit prices. Transit prices correlate positively with the shipped quantity and the negotiation power of the exporter, in this case Gazprom. Figure 4.10 depicts how the average electricity price in Europe increases with the distance of the electricity market from Russian gas.

4.6.1.2 Gas Transportation as LNG

LNG is natural gas that is compressed and cooled down to about –160°C and thereby liquefied. The volume is thereby reduced by a factor of 560–680. LNG is then transported with specially equipped deep-sea vessels. At the destination, the LNG is decompressed and re-gasified. The following is required for a complete LNG chain (Cayrade 2004; BGR 2006, p. 25):

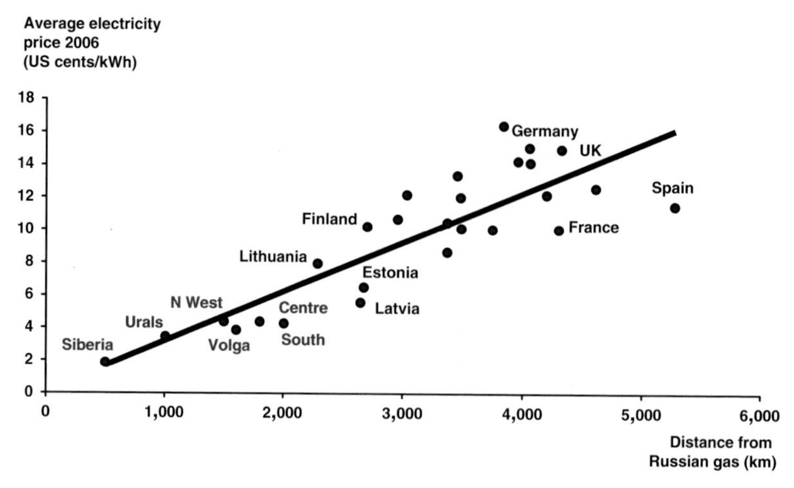

Fig. 4.10 Average electricity price and distance from Russian gas (2006) (Source: Weaving 2006)

- Liquefaction facilities; about US $1 billion for a 4.8 Gm^3 per annum facility;
- LNG tanks and LNG loading facilities;
- LNG vessels; US $300–400 million for Capesize capacity;
- LNG discharging facilities and more LNG tanks; and
- LNG re-gasifying facilities; US $300–500 million to handle 5 Gm^3 per annum.

Total cost estimates vary, depending on size, from US $2 to 10 billion for the entire chain without actually producing the gas and transporting it to the terminals. It is assumed that the chain requires 15–30% of the energy content of the LNG to run. LNG is, however, becoming more cost-efficient compared to pipeline natural gas (PNG) for distances above 2,000–3,000 km (such figures cannot be taken entirely at face value as vessels have to travel further than pipelines, which are usually laid straight across land).

The first liquefied gas was shipped from Algeria to the United Kingdom in 1964. Since then, the use of LNG has steadily increased. BGR (2006) estimates that 24%, or 211 Gm^3, of the global cross-border gas trade totaling 855 Gm^3 was shipped as LNG in 2006. Only 13 countries currently participate in this trade, Qatar being the largest exporter (15% of total traded volume) and Japan being the largest importer (39% of total traded volume). In fact, the Asian gas market is almost purely an LNG market, with a 64% LNG share; Europe is also above average with a 27% LNG share. LNG also allows gas to be shipped independently of pipelines to smaller consuming regions and from small producing regions, and offers significantly greater flexibility. For that reason the IEA estimates that LNG will reach a share of 50% of all traded gas products by 2030.

It is generally believed that coal and gas prices have little co-integration. Gas prices do, however, correlate with oil prices and many scientific papers touch on this subject. From a market point of view, it makes more sense that coal and gas prices correlate, since both are used to generate the same output: electricity.

As mentioned above, gas is the closest substitute for coal-based electricity generation. I will compare the two in more detail in Section 4.6.5.

4.6.2 Nuclear Energy

Nuclear energy today plays an important role in our electricity production. In 2005, 15% of global electricity was generated in nuclear power plants. Nuclear power accounts for approximately 6% of primary energy (see Fig. 3.1 and Appendix B). Since nuclear power production is virtually CO_2-neutral and is more economical for base load (Fig. 4.11), it would certainly be the technology for the future, were it not for the risks involved.

Before talking about the risks of nuclear energy I will summarize its history. Nuclear energy results from the splitting (fission) or merging (fusion) of the nuclei of atoms. The conversion of nuclear mass to energy is consistent with the mass–energy equivalence formula $\Delta E = \Delta mc^2$, in which ΔE equals energy release, Δm equals mass defect, and c equals the speed of light in vacuum. In the early twentieth century Albert Einstein discovered that mass can be transferred to energy. In 1938, the German chemists Otto Hahn and Fritz Strassmann proved that nuclear fission is technically feasible. In 1945, the US Army exploded the first atomic bombs over Hiroshima and Nagasaki. Since the 1950s, nuclear energy has also been harnessed for peaceful purposes: to generate electricity. The use of nuclear energy for electricity generation expanded rapidly after the oil price shocks of the 1970s. However, two major incidents shattered confidence in nuclear energy: the accident at the Three

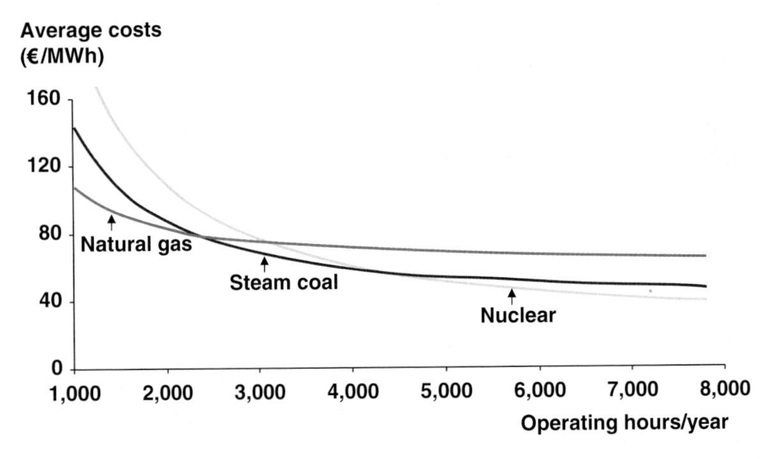

Fig. 4.11 Exemplary average costs: gas vs. coal vs. nuclear (Source: Erdmann and Zweifel 2008, p. 301)

Mile Island reactor in Harrisburg, United States, in 1979 and the Chernobyl catastrophe in Ukraine in 1986. After that, no new nuclear power plants were built outside of France and Japan until 2005 (see Erdmann and Zweifel 2008, 270ff).

Today, France has the biggest nuclear electricity share in its energy mix, totaling a staggering 79% in 2005, followed by South Korea with 37%, Japan with 28%, and Germany with a 27% electricity share. India and China, on the other hand, generate only 2% of their electricity using nuclear energy (IEA-Statistics 2005; Endres 2008). Thus, we can easily predict where the nuclear energy growth will be centered: in India and China. The safety risks involved with nuclear energy, however, will politically stall development of new nuclear power plants. Leaving aside political and strategic games on the international diplomatic playing field, it is undeniably true that by their nature nuclear power plants can cause drastic catastrophes if something goes wrong. The reason for accidents or catastrophes may include terrorism, human error, or any other cause. Another major environmental concern in relation to nuclear energy is how to deal with nuclear waste, which remains radioactive, and therefore hazardous to life, for tens and even hundreds of thousands of years.

The risks involved with nuclear energy are so large that one can argue in the interest of human survival that no nuclear power plants whatsoever should be built or operated. Germany followed this argument when it committed itself to decommissioning all of its existing nuclear power plants. It is clear that the three briefly discussed risk categories – (1) accidents in nuclear power plants, (2) disposal of radioactive waste, and (3) potential misuse of nuclear fuel – cannot be carried by any one company or risk insurer. Risks 1 and 3 cannot even be quantified and the timeframes involved can go far beyond a human lifetime. Therefore, only laws and international regulation or collaboration can reduce and handle such risks, if at all. I personally doubt that this is possible and therefore predict that nuclear energy will not be the primary choice in the future, yet will remain part of the mix.

4.6.2.1 Fuelling Nuclear Power Plants

The uranium isotope uranium-235 (^{235}U) is the most commonly used nuclear fuel. Naturally occurring uranium ore consists of 0.7% ^{235}U and 99.3% ^{238}U. The uranium isotope ^{235}U can be split by bombarding it with neutrons of lower energy. The results of this nuclear fission are 3.2×10^{-11} J of thermal heat and two to three new neutrons that can split more ^{235}U atomic cores resulting in the well-known and often-feared nuclear chain reaction.

BGR (2006, p. 23) estimates that nuclear power plants generated 390 GW_e (gigawatts of electrical power) in 2006. The world's nuclear power plants thus required about 66,500 tons of uranium ore, of which about 40,000 tons (60%) came from normal uranium ore production. The other 40% came from earlier and sometimes quite old civil inventory and, most importantly, from strategic military inventory. As such, the production of uranium ore has been lagging behind consumption for many years.

Australia, Kazakhstan, and Russia account for 40% of global uranium production. The biggest importers, the United States, France, Japan, Germany, and the

United Kingdom, have virtually no domestic production. Total uranium ore reserves are currently estimated at 1.9 Mt (27 Gt/tce); thus, at current production levels of 0.04 Mt (or 0.6 Gt/tce) per annum current global reserves will last for 48 years (BGR 2006).

4.6.3 Oil Production and Power Generation

Oil production has been studied in great detail. Therefore, I will attempt to summarize only some key elements of oil relevant for electricity generation and this book.

Here are some key facts about oil (BGR 2006; Erdmann and Zweifel 2008; IEA in general).

- Oil accounts for 35% of primary energy but only 7% of electricity generation. Thus, oil is mostly used as a fuel for transportation, rather than as a fuel for electricity generation.
- Some countries, such as Indonesia (32% in 2005), Mexico (29%), Morocco (26%), and Japan (11%), rely heavily on oil for electricity generation. However, they are continuously reducing this share, switching to coal and/or gas.
- Cumulative oil production (147 Gt) almost equals currently known oil reserves (168 Gt) (Fig. 4.12). If we include currently known resources (82 Gt) in the equation, experts expect that the 'depletion mid-point' (see also peak oil theory), where cumulative production equals the remaining potential, will be reached within the next 10–15 years. However, experts also argue that innovation and

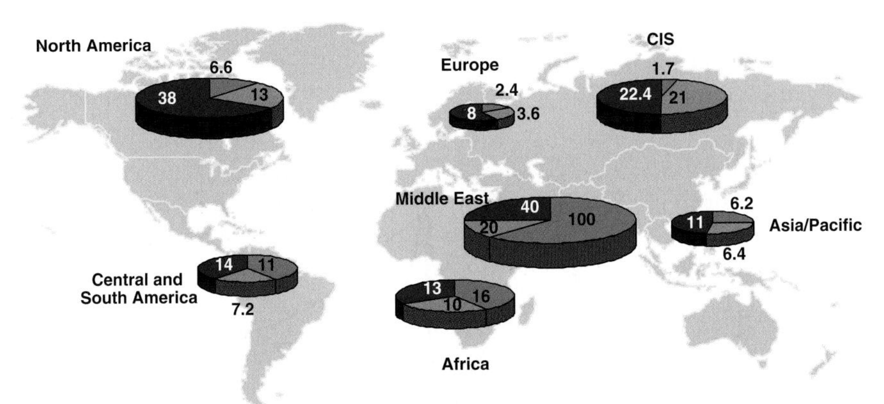

Fig. 4.12 Global oil production, reserves, and resources (Source: BGR 2006; Author's research and analysis)

newly discovered resources will further extend the depletion midpoint into the future (see Fig. 4.13).

- Current reserve/production ratio translates to 163 Gt/3.9 Gt = 42 years.
- The share of oil produced in OPEC countries is expected to further increase from approximately 41% currently. More than half of the remaining potential (reserves plus resources) is situated within the OPEC countries.
- Unconventional oil such as heavy raw oil, oil sands, bitumen, tar, and shale oil may fill part of the gap, but this is difficult to predict (see Fig. 4.13).
- Adelman and Watkins (2008) showed that the non-OPEC marginal cost of production was around US $24/barrel in 2005, while the OPEC marginal cost of production was below US $10/barrel. Other industry experts today also still assume single-digit marginal costs for OPEC oil production.

Not surprisingly, prices for oil are driven by the fuel industry market and the oil producers. The electricity market plays only a minor role, unlike for gas and coal. With oil prices so instable (reaching above US $140/barrel in July 2008 before slumping back to US $40/barrel in November 2008), the use of oil in the power industry will decrease even faster than expected. However, Adelman and Watkins (2008) showed that not only producers but also consuming countries have an interest in high oil prices as these spur research into sources of energy.

When looking at oil versus coal it becomes clear that changes in oil consumption will only marginally impact on coal. Heavy fuel oils, which have been used in power plants, can be used in other fuel applications. The technology used for oil-fired power plants can be adapted to fire other fuels such as coal and gas. Many coal-fired power plants use oil to start up a power plant before switching to coal for fueling combustion. In 2007 and 2008, some coal market participants started to look at the oil price as a proxy for the coal price. However, apparent correlation between these two energy fuels in rising markets is misleading. Most raw material

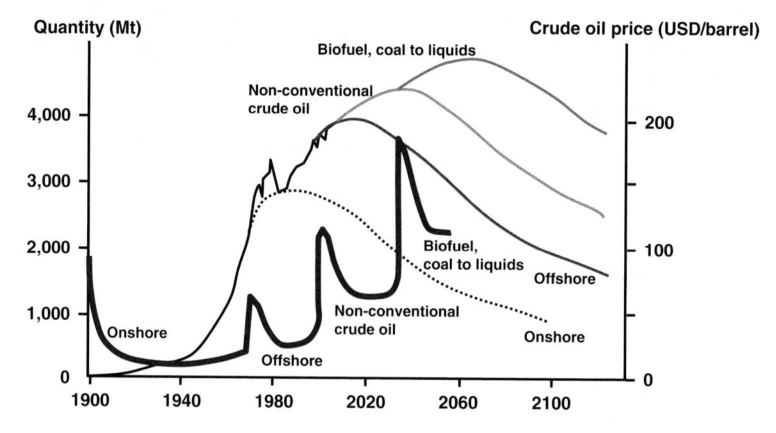

Fig. 4.13 Exemplary: perspective on future oil supply (Source: Erdmann and Zweifel 2008, p. 207)

prices, especially prices for energy raw materials, increased dramatically in 2007 and 2008. Thus, if we were to analyze this time period we would find correlation between many products that have little or no real impact on each other.

Note on correlation between oil and gas: The well-studied correlation between oil and gas prices, I believe, would not make economic sense if it weren't for the historical contractual link between the two. It may have made sense when fuel switching was still prominent, but in recent years this switching behavior has declined, or rather it has occurred in only one direction: to gas. The correlation has been well studied for many years, for instance by Brown and Yücel (2008). Hartley et al. (2008) have recently argued that the relationship between gas and oil is not as direct as previously thought.

4.6.4 Alternatives: Hydro, Wind, Biomass, Solar, and Other Sources

The alternative, nonfossil, and nonnuclear energy sources already accounted for 18% (3,300 TWh) of total electricity production (18,800 TWh) in 2005. Of the 3,300 TWh classified as deriving from 'other sources,' 2,800 TWh, or 85%, comes from hydroelectricity (IEA-Statistics 2005; BP Statistical Review 2006). Thus, non-hydro only accounted for approximately 500 TWh, or less than 3%, of global electricity production. I will discuss hydroelectricity separately, but first I want to point out some basic information about other non-hydro forms of electricity generation, as these will most likely dominate in the distant future.

The IEA estimates that the non-hydro share of global electricity generation will triple by the year 2030. Because this starts from such a small base (500 TWh) this will only cover a small portion of the increased electricity demand by that time. EIA (2008) predicts in its reference scenario that total electricity demand will increase to 33,300 TWh by 2030, thus almost doubling from 2005. The EIA estimates that by 2030 approximately 6%, or 1,000 TWh, of the increased demand can be served by non-hydro alternative fuels; the remaining 94%, or 14,980 TWh, of increased demand will have to be filled by fossil fuels, nuclear, and hydro (see Fig. 4.14).

4.6.4.1 Hydroelectricity

Hydroelectricity accounted for 15% of global electricity generation in 2005. The main hydroelectricity generators in terms of electricity share are many South American countries, Canada, China, and Russia. Also, smaller countries such as Nigeria, Switzerland, and Norway generate a large proportion of their electricity using hydro-technology (EIA 2008). Important for this study is that the EIA predicts that hydro's share will drop to 11% of world electricity generation by 2030.

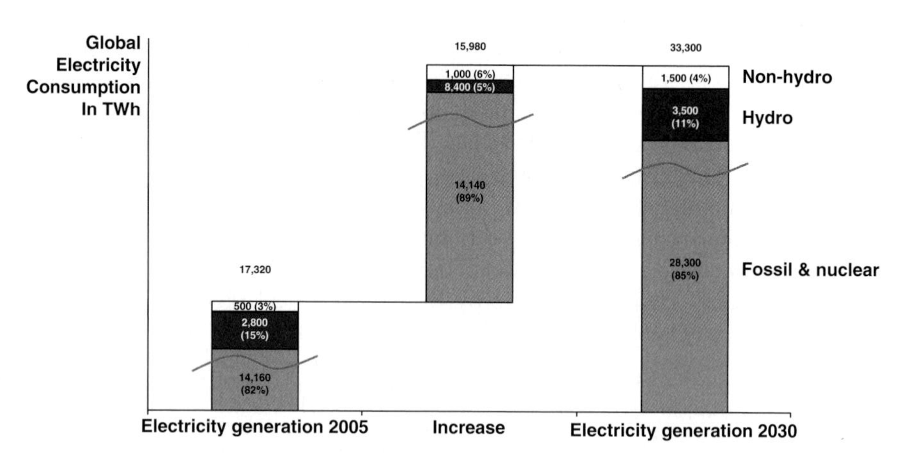

Fig. 4.14 Non-hydro and hydroelectricity growth up to 2030 (Source: EIA 2008; IEA Int'l Energy Outlook – Electricity 2007; IEA-Statistics 2005; BP Statistical Review 2006; Author's analysis)

4.6.4.2 Non-hydro Alternatives

Non-hydro alternative electricity accounted for less than 3% of global electricity generation in 2005. While they are not yet relevant on a global scale, it is important to summarize these alternatives here as they will become more prominent in the future. Non-hydro alternatives include (1) wind, (2) biogenic products, (3) solar/photovoltaic, and (4) other sources such as geothermal energy and hydrogen. We can safely assume that the 500 TWh generated by non-hydro alternatives in 2005 stem mostly from wind and maybe some biogenic products.

I personally view the non-hydro alternatives as 'solar-generated electricity.' In a way, this is where the world has to go after the 'oil age' and 'coal age' are over in perhaps 100 years. The goal is to find ways to capture the solar energy for CO_2-neutral electricity generation. However, this is less relevant for this study as its scope is merely the next 20–30 years.

Wind

Of the non-hydro alternatives, wind has developed the fastest. Some countries, such as Germany and certain other European countries, already generate a significant share (often more than 5%) of their electricity needs using wind power. The European Wind Energy Association (EWEA) estimates that 23–46 million tons of steam coal will be replaced by 20–40 GW offshore wind power by 2020 (McCloskey 2007, Issue 175, 14 December, 2007).

The problem with wind power is that much of it is still heavily subsidized and that it requires a large area to generate any significant amount of energy. Standard wind turbines today produce perhaps 2 MW of electricity; thus, 1,500 onshore wind turbines would be required to replace a standard 1,000 MW coal-fired power plant, assuming the turbines work only one-third as often and efficiently as a coal-fired power plant. Offshore wind power systems are expected to become more efficient and have less impact on the environment. The biggest problem with wind is that the wind cannot be predicted. Therefore, most countries have laws where wind power has priority over any other form of generation, often causing inefficiency with other power stations (i.e., when coal-fired power plants are switched on and off). These external factors, which also produce added inefficiency in existing power plants, are rarely, if ever, considered or quantified when talking about the effectiveness of wind power.

Biogenic Products

Biogenic products can be divided into energy plants and residual plant products such as used wood. In order to process biogenic products into gaseous or liquid fuels the industry uses (1) biochemical processes (i.e., fermentation to biogas or bio-ethanol), (2) thermochemical processes (i.e., Fischer–Tropsch synthesis used in BtL), and (3) physical–chemical processes (i.e., esterification of rape products to bio-diesel) (see Erdmann and Zweifel 2008, 254ff).

Biogenic products can support both the petroleum-based fuel industry and the coal- and gas-based electricity generation industries. Newer technology is also being developed to process biomass into coal products through biomass-to-coal (BtC) processes. These technologies promise to become more efficient, utilizing 100% of the carbon content and with 80% energy efficiency. However, the economic and energy community expects that biogenic energy products will play a small role in the future energy mix. Fuel product costs rise quickly as increasing quantities of the feed product (biomass) are required. Logistics for collecting, handling, and storing high-energy plants are a further limiting factor. Economies of scale cannot be fully utilized because not enough feed product can be sourced. Subsidies in the form of tax breaks, direct investment, or other financial support often distort the picture when comparing biogenic-based energy products to fossil-based products. However, for many farmers in the western world, growing energy crops is a welcome change from cutting down crops or reducing planted areas. On a worldwide scale, though, using agricultural land to grow energy plants seems to have a limited future as the growing population needs to be fed from a continuously decreasing agricultural acreage.

Solar, Photovoltaic, Geothermal, Tidal Power, and Other Sources

From a long-term perspective, this third category of new technologies that don't rely on wind or biogenic products seems to have the brightest future. However, today it is difficult to say if and when such non-wind and non-biogenic products/technologies will play a more important role in the planet's electricity mix.

Solar, photovoltaic, geothermal, and tidal power are currently being researched and already contribute, to some extent, to electricity needs. The use of such technologies will grow but it will take decades before they can economically satisfy a substantial share of the world's electricity needs. For example, about 3 GW of geothermal capacity was installed in the United States by December 2008; the goal is to reach 100 GW by 2050. This compares to a total globally installed capacity of 1,000 GW in 2008 (Tester 2009).

Technologies viable in the longer-term future may also include fusion power technology (see current International Thermonuclear Experimental Reactor ITER project in Cadarache, South of France, http://www.iter.org) and other technologies that use hydrogen as an energy source. Currently there are five groups of technologies that have been selected for further research as they promise large-scale hydrogen production over the next 50 years: (1) steam reforming of natural gas, (2) gasification of coal, (3) gasification of biomass, (4) electrolysis of water, and (5) thermolysis. For further details of hydrogen production technologies see European Commission (2006, 70ff). Thus, coal is also likely to be a major input for next-generation hydrogen power, though the CO_2 problem would also need to be addressed here.

4.6.5 Comparison

From the brief introductions of other methods of generating electricity above, we saw that gas is the most likely competitor for coal. When power utilities decide on new power plant projects, the question they most frequently ask is: gas or coal? The decision to build nuclear power plants, on the other hand, is usually influenced more by political than economic considerations.

Coal has the following important advantages over gas:

- larger reserve/resource base (see Fig. 4.15);
- reserves are more widely spread across the world, and there are no monopolies;

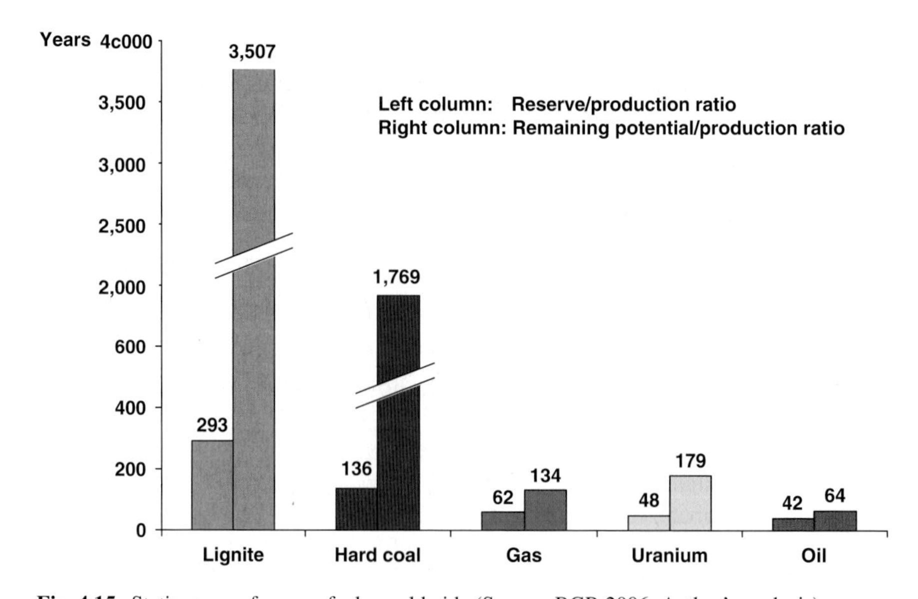

Fig. 4.15 Static range of energy fuels worldwide (Source: BGR 2006; Author's analysis)

- simpler production process resulting in low production costs;
- much easier storage and transportation from country of production to country of consumption, no pipelines or sophisticated storage equipment required; and
- lower variable electricity generation costs.

Gas has the following significant advantages over coal

- about 45% lower CO_2 emissions;
- lower investment costs in generation capacity;
- higher power plant efficiencies, thus less waste of energy resource; and
- quick turn-on and turn-off of the power plant within 5 min or less.

Because of their cost structure (see Fig. 4.16 and Fig. 4.11) in Germany and many other European countries, gas-fired power plants tend to be used for peak load. Based on the above summary of relative advantages coal appears more advantageous than gas. However, utilities need to keep their strategic energy mix. The mix is also driven by the intermediate load and peak load requirements. Gas-fired power plants can be fired up much faster than coal-fired power plants and therefore will continue to have an advantage for peak load. At the same time, decisions are often driven by politics and the concerns of surrounding populations. The main problem with gas, at least in Europe and Asia, is getting access to it and the monopolized supply structure in Russia and the Middle East. The emergence of LNG has eased that problem somewhat. Since China and India require the largest amount of new power generation capacity, and since these two countries already utilize coal for 70–80% of their generation capacity, this is where most of the world's coal-fired power plants will be constructed in the coming decades.

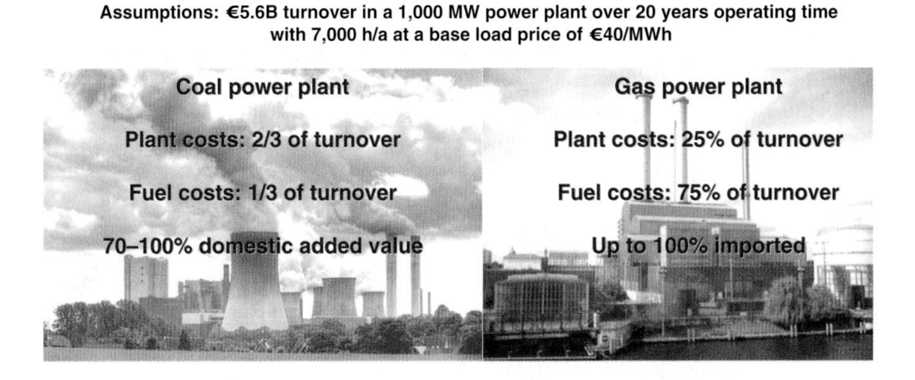

Fig. 4.16 Coal vs. gas generation costs in Europe (Source: Yexley 2006, Euracoal)

	Investment costs power generation (relative)	Operating costs power generation (relative)	Generation efficiency	Resource production/transport costs (relative	Reserves and longevity of resource	Market environment for importers of resource	Proven technology	Environmental impact	Risk of terrorism/ catastrophe	Electricity solution for next 30 years	Electricity solution in 100 years
Coal	High	Medium (€14/MWh)	Medium	Low	Very long	Very good	Yes, high	High	None	🙂 🙂	😐
PNG	Medium	High (€24/MWh)	High	Medium/ low	Not long	Not good	Yes, high	Medium	Low	🙂	🙁
LNG	Medium	High (€24/MWh)	High	Medium	Not long	Medium	Yes, medium	Medium	Low	🙂	🙁
Oil	Medium	Medium	Medium	Medium	Not long	Medium	Yes, high	Medium	Low	🙁	🙁
Nuclear	Very high	Low (€7/MWh)	High	Low	Long	Medium	Yes, medium	Low	High	😐	😐
Hydro	Very high	Vey low	High	Low	Unlimited	N.A.	Yes, medium	None	None	😐	😐
Wind	High	Very low	High	None	Unlimited	N.A.	Yes, low	Low	None	😐	😐
Other	?	?	?	?	Unlimited	N.A.	No	Low/none	?	🙁	🙂 🙂

Fig. 4.17 Qualitative comparison of electricity generation methods (Note: Operating costs are for Germany in 2005, coal = hard coal. Source: Author's analysis; Operating costs from Ernst & Young, 2006)

4.6.5.1 Long-Term Conclusion

Figure 4.17 rates the currently used electricity generation methods. Based on this table, I would rate coal as the best electricity solution for the next three decades. The next most likely energy source is gas. All other methods have disadvantages that I do not see being overcome in the next three decades. The biggest disadvantage of coal – high carbon dioxide emissions – will, of course, need to be mitigated as much as possible. Engineers and other technology experts are working on further improving efficiencies of power plants and developing methods for carbon capture and storage. Since coal also has the potential to replace oil products, I therefore refer to the coming decades as the 'new coal age,' which will take over from the 'oil age' (see Section 4.7 for more details).

In the longer term (i.e., in the next 100 years), I believe that other methods of generating electricity, such as those discussed in Section 4.6.4.2, will take over from coal. In this future 'solar age,' mankind will hopefully have dealt with the CO_2 emission problems of fossil fuels by finding safe, economic, and large-scale alternative methods of generating electricity.

4.7 The Future of Coal Use: CtL and Coal Bed Methane

I have argued that coal will remain the most important energy resource for electricity generation and is expected to increase its share beyond 40%. I have also compared coal to gas and other methods of power generation. There are a number of new and not-so-new technologies that will further increase the efficiency of coal products in today's energy-hungry economies. First and foremost is the quest to further increase coal-fired power plant efficiencies. Second, carbon capture and storage is being explored to reduce CO_2 emissions into the atmosphere (not only CO_2 resulting from coal combustion). New trends in power plant technology and CCS have already been discussed in Section 4.4.2.

In this section I will discuss two entirely different technologies outside of the current standard thermal use of coal that allow more efficient and environmentally sustainable use of hard coal and subbituminous coals. Coal-to-liquid technology is one such example that expands the use of coal into fuel products, thus competing with oil. Another example is the efficient use of coal bed methane.

4.7.1 Coal to Liquid

The quest to liquefy coal into fuel products started as early as in the first half of the twentieth century. In 1913, Fritz Bergius patented the direct hydration of coal. In 1925, Franz Fischer and Hans Tropsch patented the indirect liquefaction method, which is still referred to as Fischer–Tropsch synthesis. Germany started using liquefied coal for strategic reasons before and during World War II. Toward the end of the war Germany operated 27 CtL facilities, 9 Fischer–Tropsch indirect plants, and 18 direct liquefaction plants. By the end of the war, 90% of Germany's fuel demand was being met by CtL. After the war the technology fell out of favor, as a result of a different energy policy and low oil prices (Deutsche Bank 2007). South Africa picked up the technology to develop it further. Sasol, today's leading CtL company, converts over 43 million tons of coal into liquids using CtL technology and satisfies about 60% of South Africa's domestic fuel demand in this way, thus replacing oil imports. In the new millennium, large coal producers such as Russia's largest producer SUEK, China's largest producer Shenhua, and Australia's fourth-largest producer Anglo American have started to invest more heavily in exploring and adopting CtL technology for their difficult-to-access coal fields (see China Coal Monthly 2006; Deutsche UFG 2006). CtL is, of course, more economical when oil prices are high.

Currently, there is a tendency toward the more energy-efficient direct method of coal liquefaction, but both methods have been employed commercially (see Fig. 4.18). The main disadvantages of both CtL methods relate to the environment. CtL fuels cause significantly higher CO_2 emissions than standard fuels derived from oil. IEA CIAB-CtL (2006, p. 25) estimates that CtL transport fuel emits 2–2.5 times more CO_2 per km than standard fuels. Some experts argue that this figure is rather in the range 7–10 times. CtL plants also require 10–18 tons of water per ton of output (China Coal Monthly 2006).

Table 4.7 summarizes the current advantages and disadvantages of both the indirect and direct CtL methods.

CtL opens up a whole new era for the use of coal. IEA CIAB (2005) estimates that CtL is competitive with oil prices above US $40–50/barrel. In 2008,

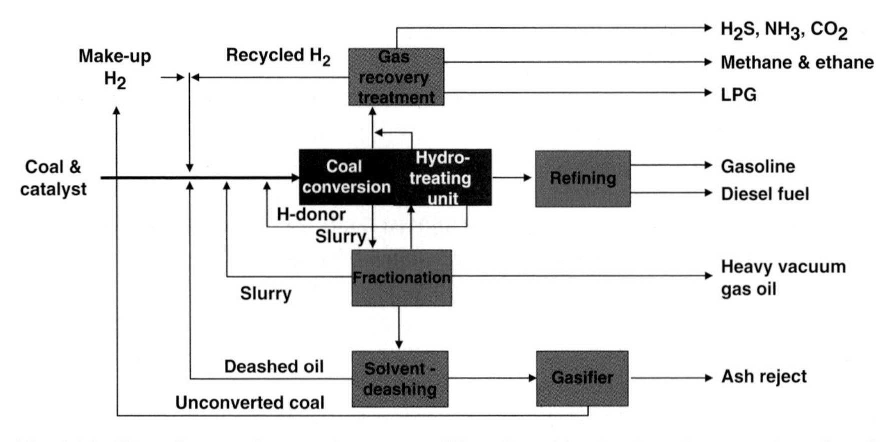

Fig. 4.18 CtL – direct coal conversion process (Note: De-ashing involves the processing of used oil into a de-ashed oil product which is not a hazardous waste but is used as a fuel. Source: IEA CIAB-CtL (2006), Workshop report; Author's analysis)

Table 4.7 CtL – comparison of direct and indirect conversion processes

Indirect liquefaction	Direct liquefaction
• Based on gasification • Converts synthetic gas (H_2 and CO) into clean methanol or hydrocarbon liquids • Can also produce ultraclean diesel or jet fuel • CO_2 can be captured for sequestration • Can coproduce electric power or hydrogen • Used by Sasol in South Africa	• Based on high-pressure dissolution and hydrogenation of coal • More energy-efficient than indirect liquefaction • Produces high-energy density fuels (diesel with low Cetane # and high aromatics) • Used by Germany in World War II, improved by the United States, now being developed in China

Source: IEA CIAB-CtL (2006), Workshop report; Author's analysis

oil prices reached well over US $100/barrel. Thus, CtL received a boost. When oil prices remain high, the implications for the international coal trade market will be substantial. While CtL fuels will not be part of the analyzed international seaborne steam coal trade, CtL producers will begin competing with power plant consumers for coal. Particularly in the next decade it is expected that this competition will be rather friendly, as CtL producers will focus on investing in CtL capacities in distant regions where it is uneconomical to mine for steam coal exports and for lower quality resources (i.e., subbituminous coal) that is also unsuitable for export in today's world. China already plans on CtL fuel accounting for 10–15% of the country's 450 million ton fuel demand by 2020 (IEA CIAB-CtL 2006). The CO_2 emission problem with CtL is a key technological aspect that needs resolving.

4.7.2 Coal Bed Methane

Coal bed methane projects are supported under the Kyoto Protocol. Through the so-called joint implementation projects investors can claim CO_2 credits, either using them to offset their own emissions or selling them on the open market. The idea is to capture otherwise wasted methane that is released (a) from abandoned underground coal fields, (b) from operating underground coal fields, or (c) from bore holes on the surface. A combined heat and power station can be fueled in this way. A case study in China presented by Faizoullina (2006) shows that one such project capturing 15 million cbm of methane per annum can supply 10,000 houses with heat while generating electricity at the same time. In addition, the project accumulates 400,000 tons of CO_2 credits per annum. For more information on coal bed methane please refer to industry sources such as http://www.cbmdata.com/.

Having looked at sources of coal in Chapter 3, and reviewed coal as a resource, analyzing the power markets and the application of coal in Chapter 4, I now turn to the global steam coal market in Chapter 5.

Chapter 5
The Global Steam Coal Market and Supply Curve

5.1 Geopolitical and Policy Environment

Einstein (1955) once said, 'Only two things are infinite, the universe and human stupidity, and I'm not sure about the former.' He might have been inspired to say this by activists who shout, 'save the planet;' as if humankind, having been in existence for only 0.04% of the planet's 5 billion years of existence, could do anything to save the planet. A more appropriate call would be to 'save humankind.' I believe that the planet will recover and go on as it has for the past billions of years even if humankind may cease to exist.

But this book is not a philosophical treatise on human existence, rather it is about coal and therefore energy, or more specifically electricity. Humans have no doubt done to the atmosphere and the planet's environment what no other living creature has done in the past; however, in order to find solutions for the future, humans will need electricity. And coal, as we have seen, is the primary source of electricity and will continue to be so for many decades to come. Unfortunately, coal is also a main contributor to greenhouse gas emissions. Therefore, governments have a responsibility to create policy that enables safe and sustainable coal use.

5.1.1 Introduction

It is the task of governments and the global community to find a framework and to decide on policy that guides the generation on the use of energy and electricity. It is also their task to regulate monopolies. It is clear that no one single government can do much about the global population problem, energy crises, and environmental issues on its own. These problems can only be solved if every nation participates. Western Europe may cut its CO_2 output by 50%, but that will be close to useless if China and India increase their CO_2 output by 10 times the amount saved in Western Europe. At the same time, it will be the utilities – the main coal consumers – that can drive many technological and environmentally sensible changes. The large multinational energy corporations will not be able to avoid their growing responsibility for minimizing the effect of their choice of generation mix on the environment and for the development and use of modern technology.

L. Schernikau, *Economics of the International Coal Trade*,
DOI 10.1007/978-90-481-9240-3_5, © Springer Science+Business Media B.V. 2010

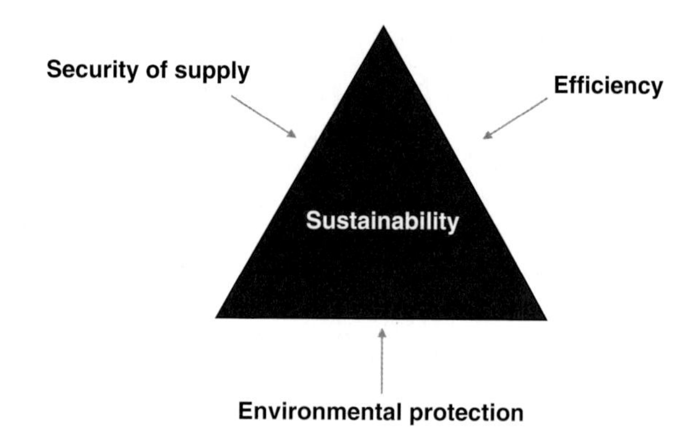

Fig. 5.1 Triangle of objectives in energy policy (Source: Author's research)

The triangle of objectives as depicted in Fig. 5.1 summarizes the main objectives of energy policy. The same imperatives apply to the utilities. It is all about sustainability (financial and environmental), driven by security of supply, efficiency, and environmental protection. What has been neglected in much of the past decade's environmental and sustainability debates is the application of sound economics in the face of enormous uncertainties when dealing with the environmental impacts of human activity in general and, more specifically, the use of coal. The application of the simple Pareto efficiency concept could not only have prevented many misguided decisions that caused major environmental damage but also have saved large amounts of money.

5.1.2 Greenhouse Gas, Kyoto, and CO_2 Trading

Greenhouse gases (GHGs) are harmful to the Earth's ozone layer and as a result can lead to increasing average temperatures, which in turn are likely to lead to climate change. At the World Climate Summit in Kyoto in 2005, 55 industrialized nations ratified the 1997 Protocol, including Annex I, and committed to reducing GHG emissions, first during the period 2008–2012. Kyoto covers the six main GHGs: carbon dioxide (CO_2), methane (CH_4), nitrous oxide (N_2O), hydrofluorocarbons (HFCs), perfluorocarbons (PFCs), and sulfur hexafluoride (SF_6). All gases are expressed as CO_2 equivalents, thus producing a single GHG reduction target.

The well-intended Kyoto Protocol was not, however, ratified by the most important nations when it comes to population, energy, and raw material use: the United States, China, and India. Vahlenkamp and McKinsey (2006) concluded that coal burning should decrease in Europe as a result of GHG-reduction policies, but will increase in the United States, where it makes economic sense without the same GHG policy. As a result, the author and Erdmann (2007) agree that Kyoto, as it is, has only a limited environmental effect, if any. In fact, it may result in higher emissions of GHGs. To illustrate this point, consider the following two scenarios that may result from the emissions trading scheme introduced in Europe in 2005 to comply with Kyoto:

1. Higher CO_2 prices in Europe could result in an avoidance strategy. But what if (as is most certainly the case) the regions to which the CO_2-emitting activity is moved produces energy with much less efficiency than in Europe? The result would be an increase in global CO_2 emissions, which was certainly not the intention of Kyoto.
2. Higher CO_2 prices will, relatively speaking, push coal prices down and gas prices up, thus increasing the spread between gas and coal. Countries with no CO_2 avoidance obligations can now buy coal more cheaply than gas; thus, they are incentivized to use CO_2-emitting coal rather than cleaner gas. As a result, global CO_2 emissions increase. This, again, was certainly not the intention of Kyoto.

Thus we can see that GHG policy makes sense only when all or at least the largest nations participate. International protocols can be useless or even harmful when nations such as the United States (22% of global CO_2 output), China (18% of global CO_2 output), and India (currently just 4% of CO_2 output) do not participate. By comparison, Germany contributes only 3% and the United Kingdom 2% of global CO_2 output (see IEA-CO_2 2006 and Appendix C).

In fact, the current Kyoto Protocol will support the increased global use of coal rather than reduce it. I strongly suggest finding every energy alternative to coal, which accounts for 40% of all anthropogenic CO_2 emissions. However, I believe that there is no way around coal in the medium-term future, and no matter what international policies are adopted, they will not prevent the relative coal burn from increasing for at least the next three decades. They can, however, influence or reduce the level of this increase and the impact of its GHG emissions.

5.1.3 Political Environment

The political environment for coal in the industrialized world is not what I would call supportive. On first glance, this may seem predictable or even justified. However, politicians – at least in Germany – sometimes lack sound economic judgment, more often due to a lack of information or understanding than to unwillingness. The results are missteps such as those described in the previous chapter. The problem with politics in a democracy is that politicians need to be elected every 4–5 years. I would argue that this system of democracy inherently favors short-term popular measures over long-term ones, initially unpopular approaches. I am not arguing against democracy, but rather for an embrace of the economic realities. There certainly are a number of politicians with very long-term views and sound analyses of the current situation, but inherently, political systems – at least in Germany – are not well equipped to deal with long-term environmental and energy security issues.

Populists in Germany want to abandon nuclear power and stop all coal burning, yet together these sources currently account for about 75% of electricity generation.

Domestic policies are about optimizing the nation's development and wealth. This is true for countries such as the United States, Russia, China, Indonesia, or Middle Eastern countries. Experts argue that most countries lack a coherent energy policy (see PESD 2009). Resources are used foremostly to cover a country's domestic energy needs. But energy needs in developing countries are very large. As a result, prices for energy raw materials will tend to rise. I already mentioned the announcement by Russia's Gazprom that it aims to export more gas and use domestic coal more within Russia, which is economically a very smart decision. South Africa is facing a tremendous coal shortage, which caused economically distressing power cuts in 2008. Indonesia wants to increase its coal burn fivefold within a decade and China turned, for the first time, from a net coal exporter into a net coal importer in 2007.

Politics is also responsible for subsidies, tariffs, and quotas. These instruments are used by governments either to protect their resources from excessive export or to protect their industries from international competition that is harmful to the nation's security or long-term existence. For the global coal trade the use of such instruments has the effect of increasing price volatility and, as a result, increasing uncertainty for coal consumers. It can be expected that more coal-producing nations will use political means to protect their resources.

Frondel et al. (2007) explained why coal production subsidies that cost billions of euros every year do not make much economic or social sense. For instance, Germany cumulatively spent €158 billion (in year 2000 prices) from 1958 to 2002 on subsidizing its domestic coal production. The authors proved that, socially, environmentally, and financially, such subsidies have been harmful. What then can or should policymakers do? I do not claim to have all the answers, but the following suggestions may assist in developing a more forward-looking, economically sound, and coherent approach (see also Erdmann 2007):

- Learn more about coal and natural resource economics. Familiarize yourself with the key figures on global electricity generation and power plant efficiencies.
- Understand the costs and time involved in replacing fossil fuels, while not forgetting the long-term goal to do so (by chance, this long-term goal is also popular most of the time).
- Encourage energy efficiency and saving; after all, the West does have a 'role model' function. But be consistent and only preach what makes economic and environmental sense.
- Reconsider Kyoto and emissions trading. Until the world is united, focus on industries that cannot move to less energy-efficient locations.
- Support closing the gap between international coal and gas prices. Right now this means supporting higher coal prices (without the cost of CO_2) and lower gas prices.
- Put greater emphasis on optimizing the use of coal rather than fighting it. Clean coal technologies need to be developed much faster and given much more support. Involve large coal-producing and coal-consuming nations early on.

This section discussed only limited aspects of the global geopolitical and policy context. I focused on less-obvious aspects of the discussion rather than repeating what can be read in newspapers and magazines on a daily basis. In summary, politics and policy are all about compromises. The coal market is still a relatively unregulated arena. I expect that this market will be subject to greater regulation in the decades to come. Competitive issues will also be looked at much more closely. The attention of anticompetitive agencies is currently focused on coal consumers, such as power utilities, rather than on coal producers. But I have shown that, in fact, the production of coal is already much more consolidated than the consumption of coal. It will therefore be interesting to see what the future will bring.

5.2 Introduction to the Global Seaborne Coal Trade

Up to the late 1970s, steam coal was consumed near its site of production. If international trade occurred, it did so across green borders (i.e., in Europe or between Canada and the United States). Following the oil crises, the coal trade began to pick up. In 1980 only about 150 million tons of seaborne steam coal was traded globally (Ritschel and Schiffer 2007, p. 23). By 2006 this figure had grown to 595 million tons. This translates into a compound annual growth rate (CAGR) of 5.4%. Germany, for instance, also appeared very late on the international coal-trading scene. The VDKI reported that in 1988 Germany imported only 6.5 million tons of steam coal. This figure grew to 33 million tons in 2006 and is likely to increase to over 40 million tons as local production is phased out. Thus, while coal demand grew at a CAGR of 1.9% from 1980 to 2005 (primary energy, see Fig. 7.4) it can be deducted that trade grows at more than twice the rate of the underlying industry demand.

As with all raw material markets, the coal market is largely about logistics. Pure mining costs account for only a small fraction of total delivered costs and 40% of FOB costs. I also discuss pure mining costs in Section 5.4. We must consider that getting coal from the mine to the power plant involves the following logistical steps: (a) moving coal from the mine to the port; (b) transshipping the coal to bulk carriers; (c) shipping coal in vessels to the destination port; (d) unloading and storing the coal at the destination port; (e) moving the coal from the destination port to the power plant; and (f) storing the coal at the power plant. Even mining, as we have seen, is largely about logistics. For that reason, I will later discuss key logistical issues, such as freight. But first I will turn to the Atlantic and Pacific coal markets.

5.2.1 Atlantic vs. Pacific Coal Markets

Figure 5.2 describes the 2006 trade flows. One can see that the Atlantic market – at its core the consumption region of Europe with total imports of 191

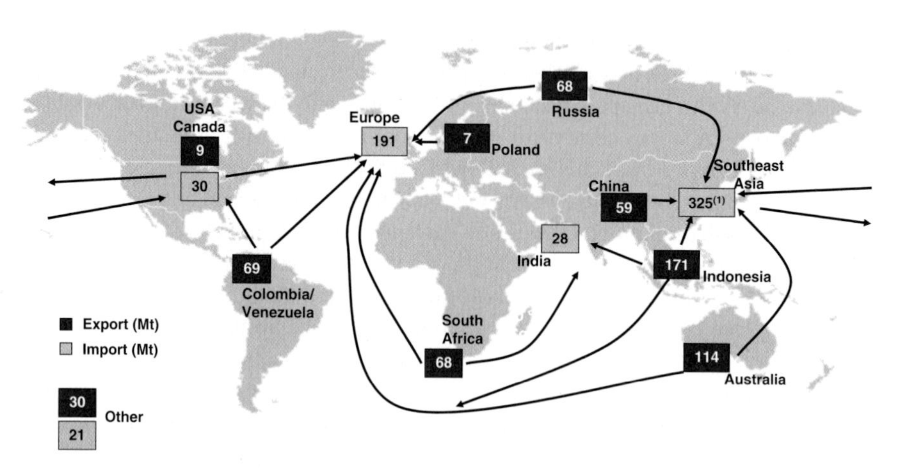

Fig. 5.2 595 million tons traded steam coal, trade flow 2006 ((1) includes China, Source: Author's research and analysis; based on VDKI 2006)

million tons – is supplied mainly by South Africa, Colombia, and Russia. As mentioned in Section 4.2.2, the United Kingdom and Germany are the largest European importers, together accounting for 70 million tons or 37% of Europe's imports. The larger Pacific market is dominated by Japan, Korea, and Taiwan, totaling 229 million tons or 70% of Pacific imports. In 2006, Asia as a whole imported 325 million tons. The Pacific market is mainly served by Indonesia and Australia (VDKI 2006).

While the Pacific and Atlantic markets have historically behaved differently, in recent years there has been more and more inter-market trade. For instance, Russian sales to Asia have increased, as has South African trade with India. Some Australian and Indonesian coal has always found its way to Europe, subject to freight viability. Figure 5.3 below depicts the market structure and how the Pacific and Atlantic markets interact.

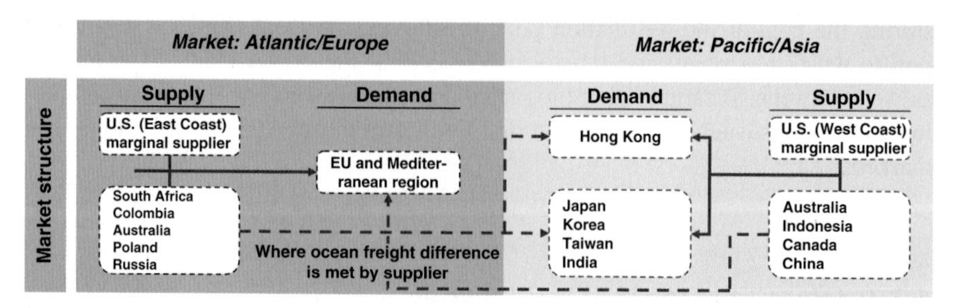

Fig. 5.3 Market structure of Atlantic and Pacific coal trade (Source: Author's research and analysis, also based on BCG 2004)

5.2.2 *Market Participants and Market Power*

Market participants in the global coal arena are of three types: (1) coal producers, (2) coal consumers, and (3) physical coal traders. These three groups greatly influence not only pricing (as discussed below) but also product flow. Producers and consumers seemingly do not require traders to do business. However, since production and consumption are continuous but only correlate on a global scale, a buffer is required.

5.2.2.1 Producers

Producers want to cover their investment and operational costs. Historically, they have a medium-term view but with coal price spikes in 2007 and 2008 they are increasingly looking for spot deals. Only the biggest producers have their own freight departments. Producers rarely act as traders. However, some large traders participate in production.

5.2.2.2 Consumers

Consumers tend to have a longer-term view driven by larger relative investment in plant and equipment. Their storage capacity is limited and therefore just-in-time delivery is becoming increasingly important, especially with higher coal prices. Many larger consumers are building up their own trading teams in order to utilize their inherent flexibility (especially where they have more than one power plant location) and to take advantage of logistical swap opportunities. Freight departments are also being built up. Japanese consumers still tend to buy mostly CIF. I would estimate that more than 50% (by volume) of European consumers are now flexible in terms of buying CIF or FOB, while the bulk of the volume is still sold to final consumers on CIF or DES terms.

5.2.2.3 Traders

Traders serve several functions in the international coal market: they (a) act as a physical buffer, (b) finance cargoes (i.e., prepay), (c) arrange freight and logistics, (d) act as outsourced purchasing or sales department, and (e) mitigate credit risk. Being a good and successful trader requires vastly different attributes to being a good and successful power generator or coal producer. Therefore, many producers or consumers utilize traders as an outsourced extension. Particularly larger utilities or producers have headcount budgets and are not as flexible in expanding or shrinking their teams. Trading, by nature, is a much more volatile business and requires flexibility that often consumers or producers, usually larger corporations, either cannot or do not want to offer. As the coal trading market turned into a seller's market in 2007/2008, more and more traders looked to secure supply by participating in coal mining and production assets.

Coal supply is being consolidated more and more as described in Section 3.5. Each supply region tends to be dominated by 3–5 producers that account for 50–80% of each region's production (also refer to Fig. 3.9). The supply market is therefore neither monopolistic nor close to perfect competition. It is fair to say that it is oligopolistic and imperfect in nature, though I have not witnessed cartel-like behavior. It is estimated that over 400 export mines supply the world coal trade. Coal demand is also being consolidated, but this development is slower and less relevant to the coal market. The demand market is far more fragmented. For example, in Europe, despite consolidation, large utilities such as E.ON, EDF, RWE, Vattenfall, Drax, and Enel account for a much smaller percentage of coal imports than do the largest producers.

There is an interesting trend that can be observed in global coal trade: exporting countries, such as Indonesia, China, Russia, and South Africa, will need to supply more and more to their own domestic coal power generators. Thus, exporting countries are being offered more competitive alternatives to pure exports. On the other hand, importing countries are relying more and more heavily on imports. This can be shown in Europe where local production is declining. This trend underlines my finding that relative coal burn will increase as more countries turn to coal generation. Overall, one can expect the market power of producers to increase.

5.2.3 Seaborne Freight

Coal is shipped in bulk carriers. Shipping not only costs money but also takes considerable time. For instance, a vessel traveling from Indonesia to Europe takes about 4 weeks and a vessel from South Africa to Europe more than 2 weeks. Thus, seaborne freight is a key CIF price determinant for coal.

Bulk carriers are divided into three categories: (1) Capesize vessels – named after the Cape because they have to ship around it – carrying 120–170 kt of bulk product, (2) Panamax vessels – so named because they are the largest vessels that can pass through the Panama Canal – carrying 60–80 kt of bulk product, and (3) handysize vessels carrying somewhere between 20 and 45 kt of bulk product. Standard routes, such as Route 4 from Richards Bay, South Africa, to ARA, Europe, are served by the largest Capesize and Panamax vessels. Less-standard routes such as within Asia or from Baltic Russia to England tend to be served by smaller handysize vessels.

Coal is only one of many bulk products carried by such vessels. Ores – here mainly iron ore – make up the largest share, followed by steam coal, then grain and then coking coal. Thus, ores, coal, and grain are the so-called major bulk products. Minor bulk products include steel, scrap, cement, fertilizers, and many others. Figure 5.4 shows the amount of major bulk products shipped with sea bulk carriers. The steam coal numbers do not add up to seaborne freight volume because this figure does not account for smaller handysize vessels or very small vessels.

The large increase in iron ore volume, which began in 2003, fueled by Chinese demand, has benefited Capesize and Panamax vessels alike. Please note that, in general, the bulk volume increase is driven by steel demand (see Fig. 5.4). Since

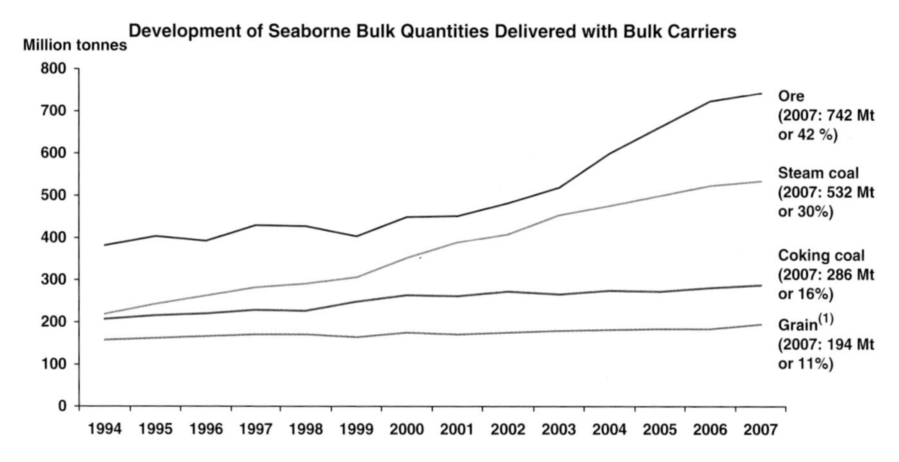

Fig. 5.4 Ore, coal, and grain determine sea freight demand, 1994–2007 ([(1)] Including Soybeans. Notes: Figures for 2006eEstimated, 2007 expected. Source: Junge 2007; Author's analysis)

coking coal is also required for steelmaking, we can easily add coking coal volumes to the general category of 'steel-related' volume, with the result that almost 60% of all freight demand is driven by steel-related bulk products. This volume increase had a direct impact on sea freight prices; while a 150,000-ton Capesize vessel traveling from Richards Bay to ARA cost on average US $7/ton for the 9-year period between 1993 and 2002, the costs rose to over US $18/ton for the 4-year period between 2003 and 2006 (Junge 2007; Galbraith's 2008). Freight prices continue to increase. Figure 5.5 shows the freight price movements for the period

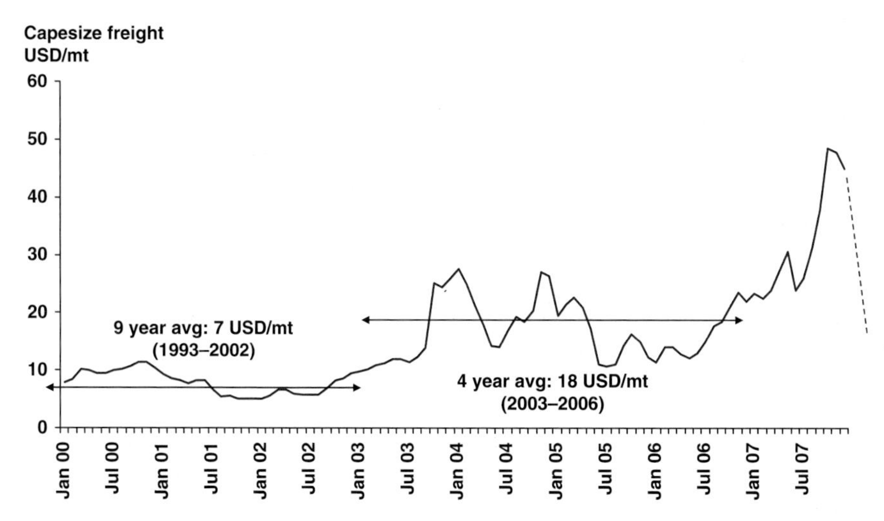

Fig. 5.5 Capesize vessel freights, 2000–2007 (Source: Author's research and analysis, Data based on Galbraith's 2008)

2000–2007. It shows that in summer 2007 a staggering cost of US $50/ton was reached, representing a 700% price increase over average prices just 5 years previously. The peak so far was reached in June 2008 when prices hit US $60/ton before falling back to around US $8/ton by October 2008. Thus, just as with commodity price movements, freight rates have reached an unprecedented volatility that market participants struggle with.

The freight market is also nothing more than a normal global market with supply and demand. Based on research, I estimate the marginal cost of running a Capesize vessel from South Africa to Europe at around US $9–14/ton in 2008. But resource (or vessel) scarcity kept prices well above that until the staggering freight price drop in fall 2008.

Freight capacity is influenced by the following basic determinants (Galbraith's 2008; Clarksons 2006; Author's Online Market Survey):

- Vessel stock (711 Capesize vessels in 2007) did not keep up with demand increase;
- New builds (600 Capesize vessels ordered by end of 2007) have about a 2-year lead time; shipyards are busy building oil and LNG tankers and container vessels;
- Scrapped vessels (only six Capesize vessels were scrapped between 2003 and 2007);
- Vessels traveling empty; in 2006 over 100 million tons of cargo moved from the Atlantic to the Pacific basin (i.e., Brazil to China), but less than 40 tons moved the other way, causing inefficiencies and more so-called 'ballasting;' and
- Asset utilization/port congestion: waiting times reaching peak levels in 2007 and 2008; anywhere between 15 and 25% of vessel capacity is usually waiting idly at ports because of congestion.

The outlook for freight rates is rather difficult. Prices halved from their peak levels in June 2008 down to around US $30/ton in September 2008 and fell even further to US $8/ton in October 2008 for a Capesize coal cargo from South Africa to Europe. Experts expect generally weak freight prices in 2009 and 2010. In the longer term, I expect that freight prices will be high enough to cover at least marginal cost while remaining highly volatile.

Freight prices made up as much as 60% of CIF European coal prices for some weeks in 2008; more recently, in autumn 2008, freight prices accounted for only about 10–15% of CIF European coal prices. Despite this large influence, there is currently little correlation between freight prices and coal prices. Coal prices have continued to increase while freight prices have decreased. But in a more balanced market environment we can expect this correlation to again become more relevant.

After better understanding the intricacies of the global seaborne coal and freight market, I will now turn to pricing in the following sections.

5.3 Contract Terms, Coal Derivatives, and Price Formation Theory

In 2003/2004, a new coal price market rally began. In hindsight it seems to have been the most staggering coal market rally in the past 100 years. Even during the oil crisis we did not see such an extreme price increase as we have seen in recent years. For example, one can see in Fig. 2.4 that the South African FOB coal price marker API4 increased from around US $28/ton in the summer of 2003 to US $168/ton in the summer of 2008 before retreating to around US $80/ton in November 2008. That means a price increase of 500% in just 5 years or almost 43% per annum

Looking back over the past 100 years, one could try to understand Fig. 5.6 and interpret the four price hikes beginning each around 1920, 1948, 1968, and 2003. However we interpret it – and the figure includes suggestions – we can see that price hikes are always followed by downturns. This is true for any market. Typically, however, in times of rising prices, nobody wants to believe in the next downturn, especially not producers, who should have considered hedging at least part of their portfolio through long-term business. Strangely enough, during times of record coal prices hardly any producer is willing to talk about forward fixed price sales longer than 6 or 12 months into the future.

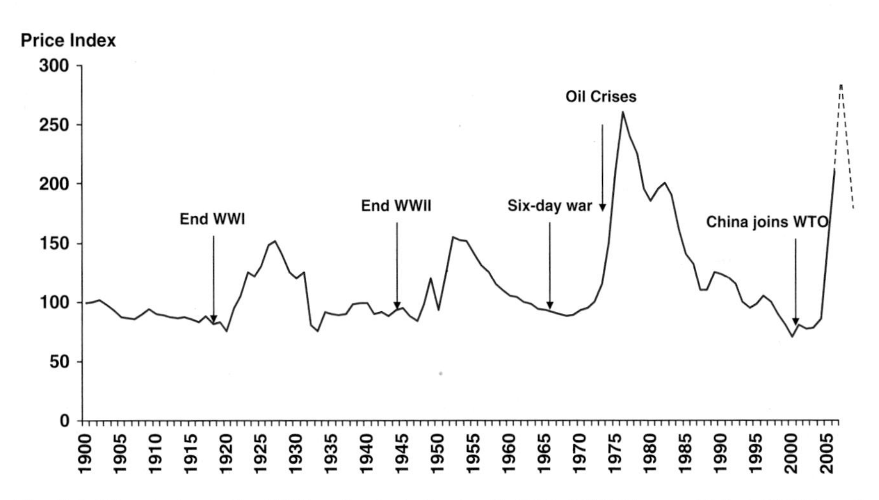

Fig. 5.6 Historic Australian FOB coal price index (real), 1901–2006 (Source: Rickets 2007 based on Reserve Bank of Australia and IEA estimates; Author's research and analysis)

5.3.1 Coal Contract Terms

This takes us to the first part of this section: contract terms. Joskow (1987) wrote the first widely publicized scientific paper about coal contract terms. His paper empirically examined data for 277 coal contracts. For each contract, Jaskow measured the duration of contractual commitments agreed to by the parties at the contract execution stage and then examined the importance of relationship-specific investments. His results support the view that buyers and sellers make longer commitments to the terms of future trade at the contract execution stage and rely less on repeated bargaining when relationship-specific investments are more important.

As already discussed in the previous section, contract term behavior has drastically changed since then. Particularly in the global steam coal trade, ever since 2003/2004 a trend toward more spot deals (3–6 months and less) can be recognized. The recent price hike, however, is only one reason for this change in behavior. Another likely reason is the emergence of paper coal derivatives. At the time Jaskow published his paper, no API2 or API4 index was yet available. Thus, floating-based pricing posed a challenge. Today, with relatively reliable coal price indices, this has become much easier. At the same time, there is less incentive to make longer-term fixed price deals, as paper coal derivatives offer an opportunity to hedge (i.e., convert floating into fixed and vice versa).

In the next section I will discuss the emergence of coal derivates and the impact these financial hedging instruments have on coal prices and the trade in general.

5.3.2 Paper Coal Derivatives and the Impact on Price and Trade

Derivatives are financial instruments that change their value in response to changes in the value or price of an underlying asset. Underlying assets can be 'common,' such as the Dow Jones index, individual company shares, oil, coal; or they may be less 'common' or more complex assets, such as loans, risks, or even other derivatives. Typical forms of derivates are forwards, futures, options, or swaps. Futures are standardized contracts traded on an exchange that require margining and therefore carry less risk. Forwards are more customized and are traded over the counter (OTC), thus bilaterally between two parties. Investment banks employ a large number of highly paid specialists to study, optimize, and create new derivatives on a daily basis. This so-called derivatives origination industry alone has become a multibillion-dollar business.

The main coal derivatives that are currently relevant for our market are standard coal forwards. Today, coal forwards can be traded OTC between bilateral parties. While shares can be traded on exchanges that standardize products and have a clearing role, this is not the case yet with coal derivatives. For any derivative to work the market requires a measure against which one can track the price of the underlying asset. Therefore, the coal market initially created two indices – the API2 index for tracking the price of CIF deliveries to Europe and the API4 index for tracking the price of FOB deliveries from South Africa. Since there is not yet an exchange that can determine the daily market price based on offers and bids, the Argus and McCloskey companies have formed an alliance and taken control of determining these indices on a weekly basis (see Argus/McCloskey Index Report). They charge a subscription fee and have established themselves as an industry standard. All coal forwards contracts are then settled based on the 'actual' API2 or API4 index published in these reports on a weekly basis. Please note that the weekly price index is determined by weekly calls made by Argus and McCloskey to selected producers, consumers, and traders to determine which deals have been done at what rate each week. Thus, a certain level of subjectivity in determining the index price level cannot be ruled out, which often concerns market participants.

The main purpose of derivatives is to reduce (or hedge) risk for one party. For instance, a coal producer may want to hedge against falling coal prices. He could then forward sell the API4 coal index, for example as a 3-month forward. This transaction does not cost him any money, except the cost of a credit margin with his bank to secure against market volatility. If the coal price falls in those 3 months, the producer could then buy back the index he has sold at a cheaper price. In this way he earns a profit on this purely financial transaction after 3 months. At the same time (i.e., after 3 months) he sells his coal at the now lower market rate. Taking the realized coal sales price together with the gain from the financial transaction, he has now secured the high market price prevailing at the time he first sold the API4 index. Similarly, but conversely, this instrument could work for consumers hedging against price increases.

Oil futures were first introduced in 1981, almost three decades ago. Since then, the oil futures trading business has overtaken the physical oil trading business by a large magnitude. Schagermann-BHP (2007) analyzed that by 1991, at the time of the first Gulf War, futures contracts tracking 1,300 barrels of oil were traded on the futures exchange for every 1 barrel that was sold physically. By 2003, at the time of the Iraq War, this ratio had reached close to 2,300:1. Today, in 2008, I assume that futures contracts tracking around 4,000 barrels of oil are traded for every one barrel of physically delivered oil, a ratio of 4,000:1.

Coal derivatives began appearing only in 1998. However, it took until 2002 to reach 2× physical volume. Figure 5.7 shows the speed at which the coal derivative volume developed. Industry experts expect the paper volume to continue to expand at almost 20–30% per year. The jump between 2006 and 2007 was most impressive, where volumes almost doubled. By the end of 2007, the multiple had

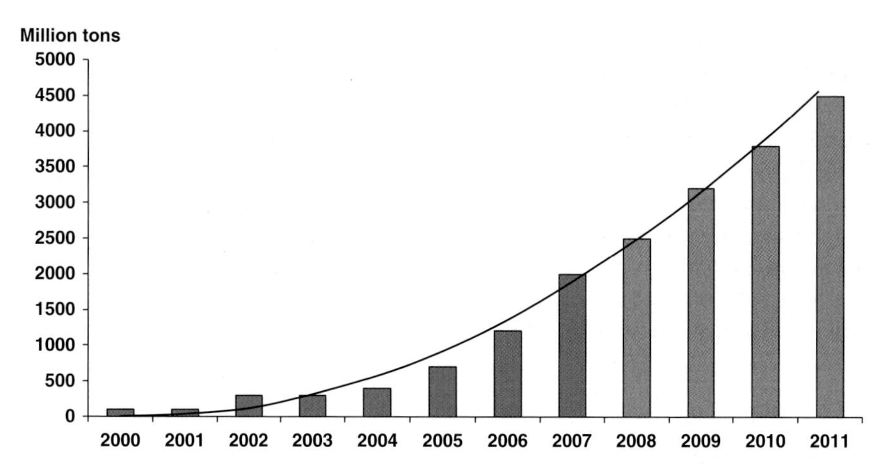

Fig. 5.7 API2 and API4 coal-derivative volume, historic, and forecast (Source: Cunningham-globalCOAL 2007; Author's analysis)

reached approximately 10× for the Atlantic market. Considering that the ratio of oil futures trading to oil delivered is 4,000:1, coal still has a long way to go. Coal derivatives have not even reached the 1984 level for oil derivatives. The next big jump will come when coal derivatives are traded with critical mass at an exchange. For instance, Germany's EEX started to include API contracts in their portfolio in 2007. By mid-2008, however, they had not yet reached critical mass but are being more widely accepted. I will briefly discuss globalCOAL and ICE, the industry's first physical exchange, in Section 7.2.

5.3.2.1 Current Problems with Derivatives

It is important to realize that derivatives are financial transactions that do not involve any physical buying or selling. Thus, paper trades are only settled financially by calculating a sum of money that is transferred from one of the two contract parties to the other. Therefore, for example, paper trade does not recognize force majeure. Neither do paper transactions allow for delivery delays because of bad weather or quality problems because the wrong coal was loaded on the vessel. While all these problems can be dealt with, often amicably, between two physically transacting parties, they are not even a point of discussion in paper trades. Therefore, when hedging physical coal positions with paper, the contracting party is aware of these 'non-hedgeable' risks. Too many physical trading parties still underestimate these risks. Hedging only works over a large number of transactions but is very risky (as strange as it may sound) on single transactions.

Another question that arises in relation to derivatives trading is whether paper prices influence physical coal prices. The answer is that they influence each other. Demand for next quarter's API2 forwards will increase paper prices, which in turn will increase the physical price at which producers are willing to sell. Conversely, when a lack of demand prompts a trader or producer to sell far below forwards prices because he is under performance pressure, this will drive down forwards prices as well.

However, a factor affecting the reliability of the API2 index remains. At least since 2006 it has been almost impossible to buy physically delivered coal in Europe at prevailing API2 prices. The FOB Richards Bay price, API4 plus spot freight to Europe was up to 11 dollars above the prevailing API2 price (see Fig. 5.8), averaging at US $4.50/ton. I have asked many market participants about this imbalance, but still have not received an adequate explanation. In general, paper hedging is about hedging price trends, not exact physical prices. But seeing how the imbalance moved US $12 in 2 months from May to July 2008, one could easily 'misjudge,' resulting in US $1,800,000 net loss, cash out, when hedging a 150,000 ton Capesize cargo and/or delivery at the 'wrong' time.

5.3.2.2 Impact on Coal Price and Trade

The impact of derivatives trading on the coal business has been tremendous. It has redefined the industry, not only changing contract term structure as discussed in the

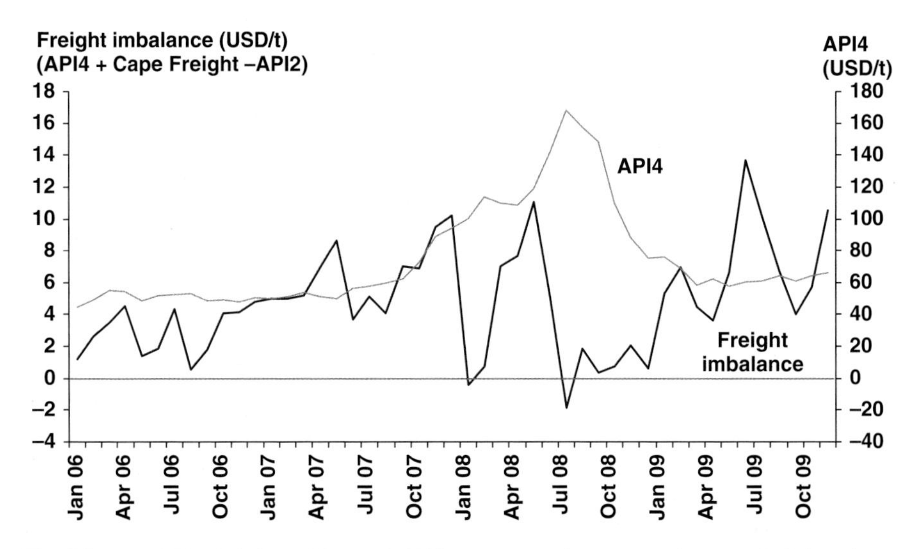

Fig. 5.8 API2 index imbalance 2006 until 2009 (Source: Author's research and analysis based on Argus/McCloskey data)

previous section but also giving the industry access to more funds and interest from many financial parties. The results may be summarized as follows:

- Bankers and fund managers have discovered coal.
- Large utility traders emerged to a large extent because they had to hedge their dark spreads anyway; thus they were the first to adopt paper trading, at the same time seeing an opportunity to develop new business.
- The coal trading market became much more transparent, making business for smaller traders more difficult.
- An influx of 'new' people into the industry made traditional multimillion-dollar handshake businesses virtually disappear.
- Price volatility increased, pushing up physical performance risk and, together with the Enron crisis, forced utilities to adopt more modern credit risk policies, which in turn reduced the pool of possible suppliers/traders.

Participants in the Online Market Survey (see Appendix D) have responded that the most likely impact of the coal derivatives market on the physical coal market is as follows (multiple answers possible, $n = 155$):

- increase in price volatility (70% of participants);
- increased influence of external markets (not directly related to coal) on coal price (58% of participants); and
- general increase in coal prices (50% of participants).

Thus, the impact of paper trading has not only been positive. The physical coal market is struggling with a new instrument that has been around only for a few years. The coming years will see market participants become more accustomed to the opportunities paper trading offers. The most positive impact of coal derivatives, in my eyes, is that it will bring greater attention to the coal business. Investment bankers today are already building up their physical coal desks. While this may scare some experienced physical traders, it will create transparency and attention. Money and the press will follow. This will allow for more investment, which will increase asset prices further, which is positive as explained in Section 5.1. But it will also make politicians and the scientific community realize the importance of coal. Maybe one day politicians will realize that there is no way around coal for at least the next 30 years, resulting in a change of attitude toward dealing with coal rather than fighting it.

5.3.3 Determinants of Coal Pricing

The 'Law of One Price' also applies to the global coal trade. Li (2008) and Warell (2007) have shown that the global steam coal market is co-integrated. As a result, the traditional price separation of the Atlantic and Pacific steam coal market is fading. Figure 5.9 depicts the standard price formation in the Atlantic and Pacific basins.

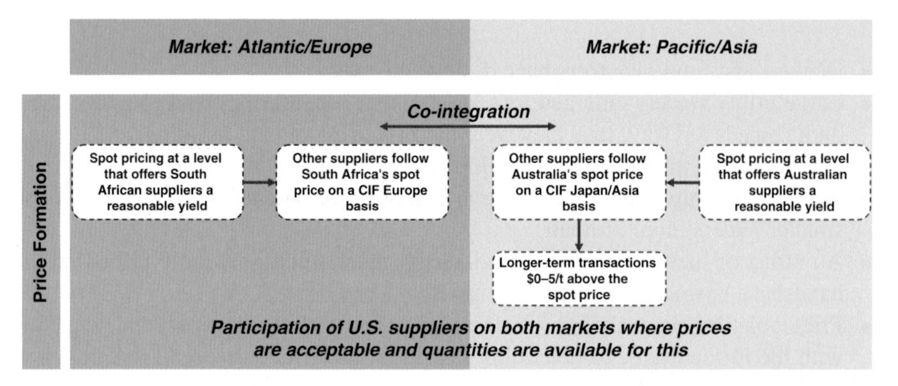

Fig. 5.9 Price formation in coal trade (Source: Author's research and analysis, based on BCG 2004)

However, pricing is still more complex than it may sound. Price discrimination can be observed in many cases. The coal market is different from most other economic markets for the following main reasons:

- Coal is a natural product and not homogenous in quality, though successful attempts have been made to commoditize at least part of the business (see globalCOAL).

- Each supply region has vastly different market conditions, often driven by infrastructure, quality, and location.
- Many destination ports are monopolized or controlled by one or two consumers. Given that not all ports can accept all vessel sizes, this leaves room for price discrimination.

While the above reasons are relevant for specific pricing at one specific time and in one specific place, I would like to continue analyzing the more generic principles of pricing trends relevant to the global coal market. I have therefore prioritized the following key determinants for coal price building (I have left out coal quality for simplification and because it is less relevant when speaking about coal price trends):

- marginal FOB costs and elasticity of supply;
- export mine capacity;
- demand growth;
- emissions prices; and
- sea transportation costs.

Mimuroto (2000) and Große et al. (2005) have also identified other determinants, including exchange rates, coal production productivity, and oil prices or, more generally, energy prices. They have shown that (1) a stronger currency of the exporting country will put upward pressure on international US dollar-denominated coal prices, (2) higher coal productivity leads to lower coal prices, and (3) increased oil prices tend to be followed by increased coal prices. Regarding oil versus coal prices, please see my comments in Section 4.6.3.

I will now discuss the above-mentioned five key determinants in more detail.

5.3.3.1 Marginal FOB Costs and Elasticity of Supply

The market price in perfect competition equals the marginal cost. In a buyer's market, disregarding political or policy factors, this would also be the price formation for coal. In fact, in the low-price years in the early 2000s this was very much the case. Real marginal FOB costs for steam coal will be discussed in more detail in Section 5.4 since these costs provide the basis for coal pricing. There, I will also discuss elasticity trends. It can be seen that the likely trend is toward less elastic supply, resulting in higher price volatility.

5.3.3.2 Export Mine Capacity and Demand Growth

When analyzing export mine capacity, the amount of coal produced for export equals supply and the amount of coal sold or shipped equals demand. Therefore demand growth and export mine capacity need to be looked at together. Coal demand growth has increased in recent years. Figure 5.10 below shows historic and future forecasted demand growth. The steam coal trade has been increasing at an average annual rate of 6–7% (Kopal 2007, p. 29), which is expected to continue,

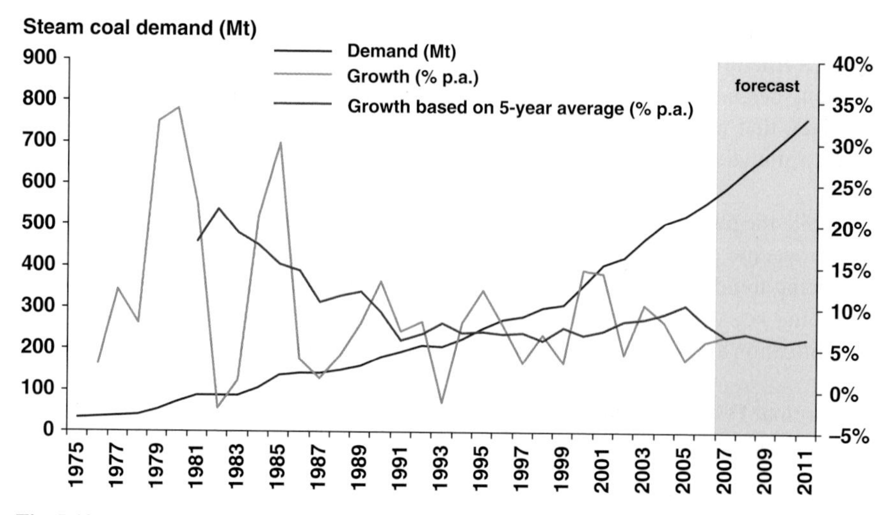

Fig. 5.10 Demand growth for seaborne traded steam coal (Source: Kopal 2007; Author's analysis)

well ahead of the projected total global coal demand increase of 2.2% per annum (see Table 4.2: IEA Forecast of Global Primary Energy Demand, 1980–2030).

Actual export mine capacity utilization of around 80% is 'healthy' for a functioning and efficient market. Kopal (2007) has analyzed historic and future export mine capacities and has determined through an extensive analysis of hundreds of individual export mines that capacity utilization has been above 90% since 2003. A capacity utilization of 100% is currently expected to be reached in 2009. In 2001 this level of scarcity was predicted for 2006, thus 5 years into the future. This timeframe has now halved to around 2.5 years into the future.

The scarcity of coal supply started in 2003, just before prices started to increase. In Fig. 5.11 I detail supply and demand and the resulting capacity utilization starting in 2002 and projected until 2011. The data from Kopal (2007) have been adjusted taking current (first half 2008) knowledge into account.

5.3.3.3 Emissions Prices

As discussed in Section 4.6, one major disadvantage of coal is the relatively higher CO_2 emissions compared to the nearest substitute, natural gas. Since power prices also follow the 'Law of One Price,' at least in each region, increased CO_2 prices have a direct effect on what utilities can pay for steam coal. The clean dark spread measure, discussed in Section 4.3.2, has been developed to account for emissions. Therefore, increased carbon dioxide prices will reduce coal prices in the medium to long term as utilities switch to natural gas. Lower coal prices relative to gas prices are not desirable from an environmental point of view. It seems counterintuitive but has been discussed in Section 5.1.2.

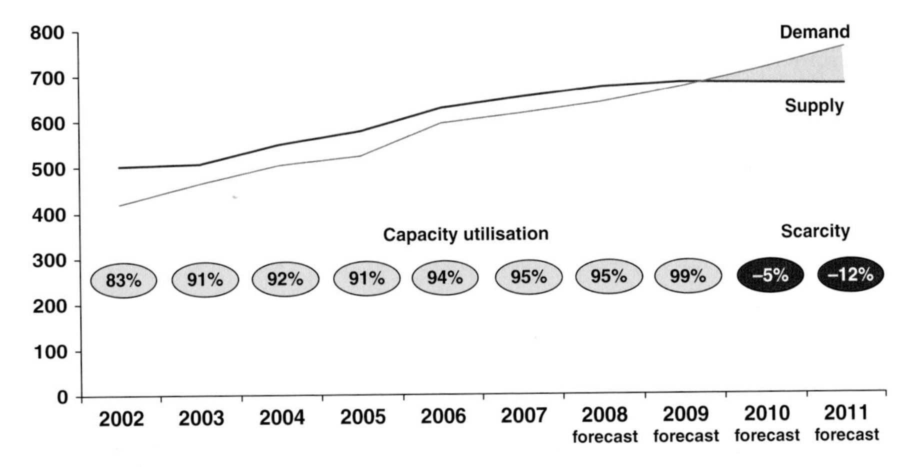

Fig. 5.11 Capacity utilization forecasts for export mines. Note: Supply assumes known mine extension and agreed new mine developments as of January 2007 (Source: Author's research and analysis, based on Kopal 2007)

5.3.3.4 Sea Transportation Costs

Sea transportation costs are a crucial element of delivered coal costs. They have been discussed in Section 5.2.3.

When analyzing the more recent price increases during 2007 and 2008, participants in the Online Coal Market Survey identified four main reasons for the price increase (see Fig. 5.12): demand in the Pacific market, oil price, freight rates,

Most important drivers for historic and future prices (increasing prices)

1. Demand in the Pacific market (Asia)
2. Oil price (correlation of energy prices)
3. Freight rates
4. Influence of market players trading coal derivates
5. Domestic coal demand in coal exporting countries
6. Capacity utilization of export ports
7. Production costs
8. Demand in the Atlantic market (Europe)
9. Utilization of global ship capacities of dry bulk carriers
10. Global coal supply
11. Supplier's price policy by price fixing
12. Capacity utilization of inland transportation
13. Capacity utilization of producing mines
14. Don't know – no answer

Future price drivers
Historic price drivers

0 1 2 3
Points (standardized)[1]

Fig. 5.12 Price drivers from online market survey ([1] Points are standardized based on average: 3 points for 1st place, 2 points for 2nd place, and 1 point for 3rd place. Note: n=102, here only participants that forecast higher future coal prices. Source: Author's online coal survey)

and coal derivative traders. When looking at future price drivers, survey participants prioritized the same top three drivers (demand in the Pacific market, oil price, and freight rates) but thought that fundamentals such as domestic coal demand, production costs, and global coal supply will become more important.

After discussing FOB marginal cost later in Section 5.4, I will analyze coal prices on a delivered basis, thus including sea transportation costs, with our GAMS-programmed computer model WorldCoal in Appendix E.

5.3.4 Economic Theory and Coal Pricing

Having discussed the real-life economic determinants of coal pricing in the previous chapter, I now turn to basic economic theory relevant to pricing. One can distinguish between price building mechanisms in four kinds of different markets that I will discuss next.

- Perfect competition: price = marginal cost.
- Monopoly: price with highest return for monopolist.
- Regulated monopoly: price = average total costs = tariff.
- Imperfect competition: strategic pricing.

5.3.4.1 Perfect Competition

In a perfectly competitive market demand equals supply and the equilibrium price equals the marginal cost of production or, applied to coal, the marginal cost of the most expensive global coal producer. Perfect competition exists when no single producer or consumer can impact the market price level or the total quantity on the market. In this case, the price is determined by the additional marginal cost of one extra unit of output. This makes logical sense: in such a market the producer would continue to produce only until each additional unit stops generating an additional profit margin. If he were to produce more, he would have a negative contribution margin, and thus not contribute to covering his fixed costs, actually losing money as a result. If he were to produce less, he would forego a positive contribution margin that would help cover his fixed costs, thus foregoing profit. Formula (5.1) below summarizes the price calculation for perfect competition. Section 6.1 will examine perfect competition as a reference case including exact definitions.

$$p = MC \tag{5.1}$$

5.3.4.2 Monopoly

In a monopoly there is only one supplier that has full market power and control over quantity and therefore price. The monopoly price is hence higher than the price for perfect competition, allowing the monopolist to maximize his total profit. He can reduce the supplied quantity until price times quantity results in the maximum profit. Formula (5.2) below summarizes the price calculation for a monopoly. In Section 6.1 I will discuss the reference case monopoly including exact definitions in more detail.

$$p = \frac{a - c_v}{2} + c_v \tag{5.2}$$

5.3.4.3 Regulated Monopoly

In a regulated monopoly, such as in a government-controlled supply situation, the regulator sets the quantity (quota) or price (tariff) in such a way that the monopolist can cover his total costs. Price therefore equals the average total cost. Formula (5.3) summarizes the price calculation for a regulated monopoly.

$$p = \frac{\mathrm{TC}}{Q} = \mathrm{ATC} \tag{5.3}$$

5.3.4.4 Imperfect Competition

Under imperfect competition the price will reach a value somewhere between the monopoly and perfect competition prices, higher than perfect competition, but lower than monopoly. Game theory offers theoretical economic models to solve such imperfect competitive conditions. I have argued in this paper that the coal market is characterized by imperfect competition. I have shown that the market is unified across the globe, but has oligopolistic tendencies. The coal product is not homogenous and prices have arguably been above the marginal cost of production, at least in 2007 and 2008. I will analyze game theory in more detail starting in Chapter 6, where I will extend game theory and Cournot to include increasing marginal cost, thus softening the constant marginal cost assumption. In the following chapters I will show that coal does not have constant marginal cost but rather variable, increasing marginal cost. I will begin with the real global FOB cost analysis for the years 2005 and 2006.

5.4 Variable Cost Analysis – Real Global FOB Costs

The international steam coal supply for export is a relatively non-transparent market. There are very few global coal supply cost analyses available today. Ritschel and Schiffer (2007), McCloskey and Baruya (2007), and Simes and Jonker (2006) are among the few who have attempted a global overview. Others have only analyzed individual markets in greater detail. Based on market knowledge I have constructed an FOB marginal cost supply curve for the years 2005 and 2006. I was able to draw from a wide range of experience, including but not limited to trading coal products of all origins other than Australia and analyzing, investing in, or partnering in coal mining projects in South Africa, Colombia, Russia, and Indonesia.

5.4.1 Methodology

The data stem from over 100 interviews with producers, traders, and consumers held in 2006 and 2007. I have also drawn on in-depth analyses of industry conference presentations and papers, scientific publications, and proprietary financial and economic documentation from industry participants.

I divided the coal supply world into eight separate supply regions: Indonesia, Australia, Russia, South Africa, Colombia/Venezuela, China, Poland, and others. I divided the suppliers of each region into sub-supply regions with varying marginal cost. Next, I categorized each individual sub-supply region by underground versus surface mining. Finally, I analyzed each sub-supply region and estimated a single value for three FOB cost components: (1) mining costs, (2) inland transportation costs, and (3) transshipment costs onto the mother vessel.

I also introduced the variables CPR, CIT, and FOB since they are used in the GAMS-programmed WorldCoal model discussed in the next chapter.

This methodology uses a number of approximations that should not influence the overall result, but limits accuracy on a sub-supply region level. In another analysis, one could analyze all export coal mines, not just supply regions. This exercise, however, would require a lot of time and expenses for travel or subscription fees as such information is not readily available. It would, however, allow us to account for variances in coal quality.

5.4.2 Results

The result is a three-dimensional matrix of FOB variable costs that I analyzed for 2005 and 2006. Figure 5.13 shows the resulting 2006 FOB marginal cost curve for globally traded steam coal. As with other raw materials, the marginal cost curve is

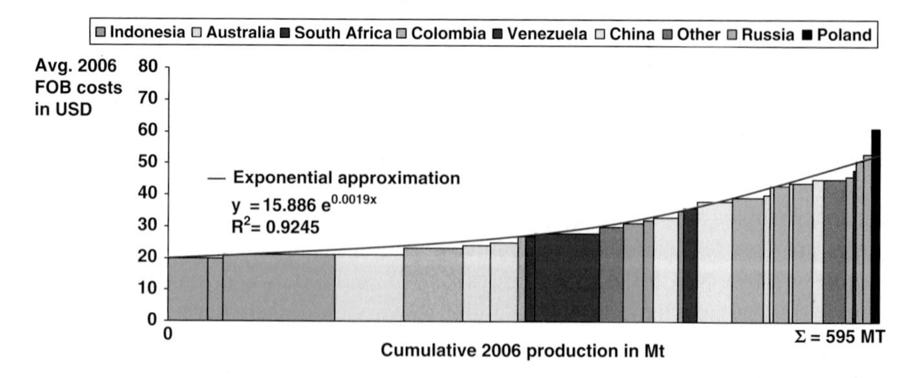

Fig. 5.13 2006 FOB supply curve by country (Source: Author's FOB marginal cost analysis)

not constant, but increasing (or variable). The reason is that the variable cost of production of raw materials increases with time as one depletes a resource, because the resource needs to be mined from deeper seams and because quality often diminishes over time.

Figure 5.13 details the supply countries. We can see that Indonesia, Australia, Colombia, and South Africa are the cheapest producers. Countries such as Russia and Poland set the price floor for this market. Please note that the average API4 (FOB South Africa) for 2006 was US $50.71/ton, thus well above South Africa's most expensive producer and in the range of marginal Russian FOB costs.

Figure 5.14 below depicts the 2005 FOB coal supply curve by comparison. This time the graph details mining, inland transportation, and transshipment costs. It is interesting to see that mining accounts on average for only about 40% of total FOB costs in 2005 and 2006. Inland transportation accounted for 44% and the approximately 16% remaining is spent on transshipment (see Fig. 3.10).

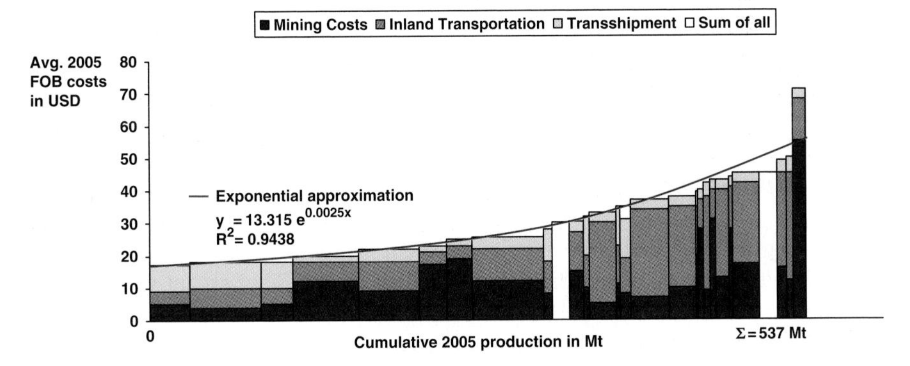

Fig. 5.14 2005 FOB supply curve by type of cost (Source: Author's FOB marginal cost analysis)

Average 2006 FOB costs were US $29/ton (2005: US $28/ton), with the marginal supplier, Poland, reaching US $61/ton. Since Poland is becoming a smaller exporter (with seaborne exports dwindling to below 5 million tons by 2008), I would estimate that 2006 global marginal FOB costs were closer to the mid-50s, driven by Russia. 2005 global marginal FOB costs were still slightly below 50, which is illustrated by the 2005 real average API4 of US $46.01/ton.

Figure 5.15 shows the 2006 FOB supply curve again, but now detailing underground versus surface mining by sub-supply region. It is not surprising to see that surface mines have on average lower marginal costs than underground mines. For 2006, the average marginal mining cost (not FOB cost) for underground mining was US $15/ton versus US $10/ton for surface mines. I also determined that underground mines are generally located further away from loading ports, resulting in average inland transportation costs of US $20/ton for underground mines versus only US $11/ton for surface mines (see Fig. 3.10). Surface mines have a slightly higher transshipment cost, which is caused by Indonesia, a surface-mining country,

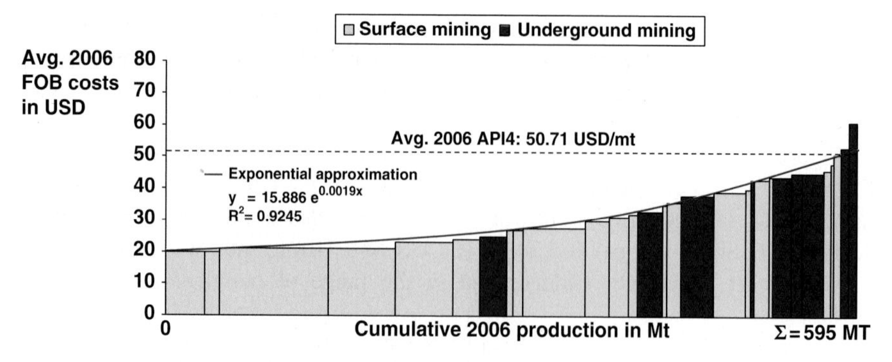

Fig. 5.15 2006 FOB supply curve by mining method (Source: Author's FOB marginal cost analysis)

that uses barge transshipment and thus increases the average. The average of all FOB marginal costs for underground mining came to US \$38/ton versus US \$26/ton for surface mines. Thus, underground mining is almost 50% more expensive than surface mining on an FOB basis.

5.4.3 Implications

Comparing the 2005 and 2006 supply curves, it can be seen that supply increased by 11% in that one year, significantly more than the 6–7% of recent years, which was also predicted for the near future (Kopal 2007, p. 29). At the same time, the supply curve has shifted upward and extended to the right (see Fig. 5.16). It appears that the curve also has flattened. This could make sense as the general cost basis increases and the relative difference between the marginal producer cost and the cheapest producer cost may decrease as overall costs increase. However, generally I would expect coal supply to become less elastic, especially when export mine capacity utilization nears full capacity.

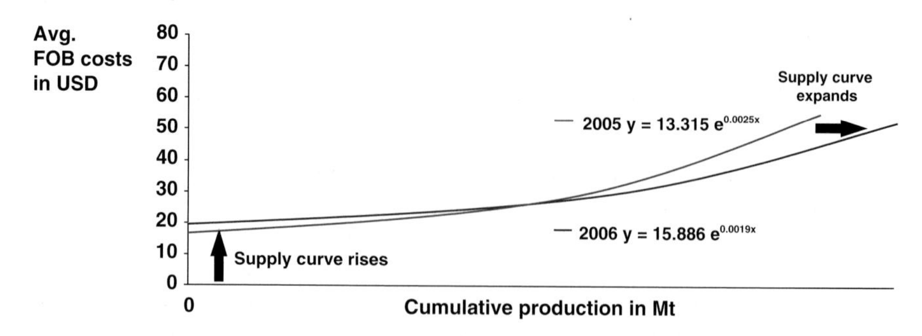

Fig. 5.16 Coal supply elasticity, 2005 vs. 2006 (Source: Author's FOB marginal cost analysis)

I have shown that the marginal costs for coal production are not constant, but in fact increasing, and thus variable. This is a phenomenon that sets coal apart from normal manufactured goods, but one that is also experienced with other raw materials. I have shown in Section 5.3 that the coal market has imperfect competition. Thus, game theory and Cournot may be used to scientifically describe the market. Before turning to game theory in Chapter 6, I will first discuss the proprietary coal market model WorldCoal programmed in GAMS in the nextsection.

5.5 WorldCoal: GAMS-Programmed Coal Market Model Including Sea Freight

What follows is a brief description of the WorldCoal model. A full description of this model is contained in Appendix E.

WorldCoal is a nonlinear model to quantitatively analyze the global steam coal trade. It enables real-time analysis of the current market situation, which was done for the year 2006 as described here. With this model, it is possible to analyze certain likely future scenarios, such as production capacity or logistical constraint scenarios. WorldCoal was programmed by Endres 2008 as part of Stefan Endres' research project, coached and supervised by the author and Professor Georg Erdmann, and the Department of Energy Systems at the Technical University in Berlin. WorldCoal models CIF cost and price levels in 2006, thus extending beyond the previously discussed FOB cost levels.

The WorldCoal model uses GAMS programming (General Algebraic Modeling System) to optimize the distribution of the 2006 coal supply in such a way that the total cost of the market (TC) is minimized. A maximum supply for each supplier was defined, based on production or logistical constraints. These constraints were calibrated with real-life constraints in the various markets for the year 2006, such as in South Africa (as discussed in previous chapters).

Modeling of the global steam coal market corresponds to solving a transportation problem. Rosenthal (2008) describes in the *trnsport.gms* model a basic market with two producers and three consumers. WorldCoal extended that model. GAMS optimizes transportation for the total cost minimum given the constraints of the producers (supmax and supmin) and demand by consumers (dem). Today there are many examples of complex models for determining energy market equilibriums. For instance, Holz et al. (2008) modeled the European gas market in their model GASMOD. When modeling the steam coal market, I utilized much of the systematic of GASMOD but adjusted appropriately for the coal market.

As already discussed in Section 5.4, supply is divided into various supply regions. The same is true for demand. In the WorldCoal model the regions were slightly adjusted compared to the FOB supply curve regions discussed in the previous section. Russia was considered as two supply regions, one for the Atlantic (west) and the other for the Pacific (east). In total, the model differentiates between eight supply regions and nine demand regions as detailed below in Table 5.1.

Table 5.1 WorldCoal model's geographic definitions

Exporters			Importers		
Country	Abbr.	Port	Country	Abbr.	Port
Australia	AU	Newcastle	Europe	EU	ARA
South Africa	SA	Richards Bay	Japan	JA	Yokohama
Indonesia	ID	Banjarmasin	Korea	KR	Ulsan
Russia-Poland West	RU_WEST	Riga	Taiwan	TW	Kaohsiung
Russia-Poland East	RU_EAST	Vostochny	China	CH_IM	Hong Kong
Colombia-Venezuela	CO	Puerto Bolivar	India	IN	Mundra
China	CH-EX	Rizhao	Latin America	LA	Acapulco
USA-Canada	US_EX	Jacksonville	USA-Canada	US_IM	Mobile
			Asia Other	AS_OT	Singapore

Sources: Endres (2008) and Pinchin-Lloyd (2005)

For the purpose of the WorldCoal model, the seaborne steam coal market is considered to be perfectly competitive. Thus, no one single supply region or demand region can influence price or quantity; rather, they have to take these as given (Varian 1999). The individual regions' supply and demand functions add up to a worldwide supply and demand function. I also assumed this principle when building the FOB supply curve in Section 5.4. This assumption is reasonable for the global supply curve, especially when taking a single supply region as a competitor. No one single supply region has any market power, just as no one single demand region has any market power. In perfect competition, the market price equals the marginal cost of the marginal supply (see also the price discussion in Section 5.3.4).

5.5.1 Implications

WorldCoal was programmed in GAMS to model the international seaborne steam coal trade market. The planning phase resulted in a nonlinear system of equations. With simplifying model restrictions, WorldCoal then calculates the optimal market equilibrium given the total-cost-minimizing objective function. The model's results for the reference year 2006 were surprisingly close to the real market. Market prices differed on average only by about 6%, despite the many simplifications, including but not limited to the single-quality assumption. This is an indication that the 2006 coal market operated close to the theoretical market equilibrium in a perfect competitive world.

Shadow prices were introduced for sensitivity analysis. The model was also run replacing the minimizing total cost (COST) objective function with a maximizing total profit objective function. Except for a small market adjustment that reduced the quantity supplied to India by 16 million tons, everything else remained constant. This is also an indication that the market seemed to be close to its optimum and there was little, if any, strategic supply shorting by the market's supplier.

The 2015 scenario restrictions were such that the supply becomes even more concentrated, with many demand regions having only one supplier. In the real coal world, this will not happen. It is unlikely that any demand region will take the strategic risk of depending on a single supply region for a longer period of time, even if this were to save money. Such strategic thoughts could not be modeled into WorldCoal, however. The 2015 scenario forecasted prices in the range of US $170–180/ton. These prices are up to three times higher than 2006 price levels. After WorldCoal programming and analysis were finished, coal prices already skyrocketed to above US $220/ton in July 2008, driven by Asian demand. Thus, WorldCoal at least forecasted the trend, also driven by Asia. The question of what will happen from here still remains, which I discuss in the final chapter.

WorldCoal only models an entire year. Thus, any intra-year volatility cannot be described, and in fact wasn't meant to be. When modeling a global market, the point is to get an understanding of what will happen when certain scenarios are realized. They are also useful for predicting long-term market trends. As such, I believe that WorldCoal is fulfilling its purpose for the global coal trade market. However, I am aware that WorldCoal has many limitations that future researchers will hopefully reduce. But as with any model, it has to remain manageable and thus will always require some simplification.

WorldCoal analyzed the coal market under the assumption of perfect competition. However, I have shown in this study that coal supply has oligopolistic tendencies. Thus, it is worth comparing the WorldCoal model to a Cournot model. While building a Cournot model is not a subject of this study, I will try to build a basis for such a model by studying Cournot in the following chapter. The goal is to calculate the Cournot equilibrium with increasing marginal cost – as is the case for the coal market and any other raw materials supply market – rather than constant marginal cost.

Chapter 6
Industrial Structure: Game Theory and Cournot

6.1 Introduction

The study of Cournot and thus game theory is essential for the coal market as it estimates how producers can influence the market price and market quantity. The coal market is not a perfect competition, nor is it a monopoly. As such the Cournot duopoly provides insight from a theoretical perspective into how the market behaves and how much market power even smaller players have.

It has been empirically shown and discussed in Section 5.4 that marginal FOB costs for steam coal are increasing exponentially rather than remaining constant. The discussion about Cournot competition with increasing marginal cost probably started in the 1950s; thus, it is not novel in itself. However, the available literature is generally of a very complex nature and not easy to find. Standard text books, such as Tirole (1988), do mention the possibility of nonlinear cost functions in an abstract form, but tend to provide examples for linear cost functions. Some older and newer literature, such as Fisher (1961) and Kopel (1996), discusses nonlinear cost functions in more detail but does not mention the impact of price and profits, as I will here.

Cournot competition is an economic game theory model used to describe industry structure. It was named after Antoine Augustin Cournot (1801–1877), based on his observations of competition in a spring water duopoly. Cournot's model contains the following features:

- There is more than one firm and all firms produce a homogeneous product;
- The firms do not cooperate;
- The firms have market power;
- The number of firms is fixed;
- The firms compete in quantities and choose quantities simultaneously;
- There is strategic behavior by the firms.

An essential assumption of this model is that each firm aims to maximize profits. The following is also assumed:

- Price is a commonly known decreasing function of total output;
- All firms know n, the total number of firms in the market;

L. Schernikau, *Economics of the International Coal Trade*,
DOI 10.1007/978-90-481-9240-3_6, © Springer Science+Business Media B.V. 2010

- Each firm has a cost function $c_i(q_i)$;
- The cost functions may be the same or different among firms;
- The market price is set at such a level that demand equals the total quantity produced by all firms.

The Cournot model has the following key implications: (1) output is greater with Cournot duopoly than with monopoly (in extreme cases it would be the same), but lower than in the case of perfect competition, and (2) price is lower with Cournot duopoly than with monopoly (in extreme cases it would be the same), but not as low as with perfect competition.

According to this model, the firms have an incentive to form a cartel, effectively turning the Cournot model into a monopoly. However, cartels are usually illegal, so firms have some motive to tacitly collude using self-imposing strategies to reduce output, which (ceteris paribus) raises the price and thus increases profits, if the price effect is greater than the output effect.

Before discussing game theory and Cournot, the following two reference models should be discussed and understood: perfect competition and monopoly. The following discussion is useful in gaining a better understanding of Cournot in general.

6.2 Reference Model 1: Perfect Competition

Perfect competition is characterized by the fact that each market player has no market power and cannot influence the price. The market price cannot be influenced because the total quantity can only be minimally influenced by any individual market player. We already briefly introduced perfect competition when discussing price theory in Section 5.3.4.

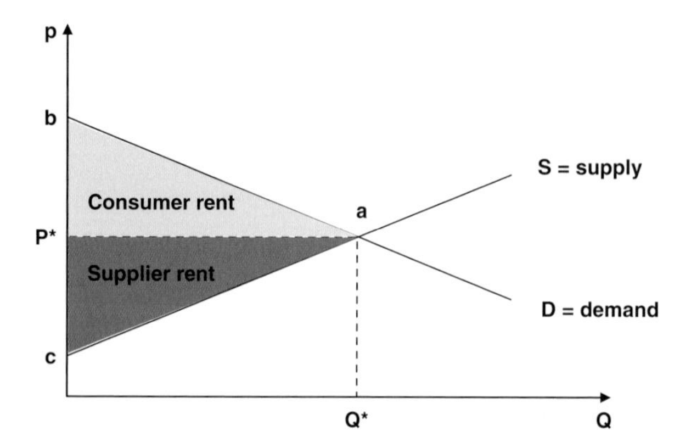

Fig. 6.1 Graphic representation of the reference model perfect competition (Source: Author)

It is important to study perfect competition in relation to the coal market as well. The key principle – that the market price equals the marginal cost of the marginal supplier – also applies to the coal market. When discussing Cournot, perfect competition is a reference model and forms the lower boundary for the price in Cournot where n is indefinite.

Figure 6.1 below shows graphically perfect competition and the fact that the market price is determined by the marginal cost of the marginal supplier to the market.

6.3 Reference Model 2: Monopoly

In a monopoly there is only one player, in our case one coal producer, that has full market power and full control of the market price by setting quantity. In perfect competition the price does not depend on the quantity of one market player, whereas in monopoly the price naturally depends on the monopolist's supplied quantity.

Monopoly is at the opposite extreme to perfect competition. Here one player has all market power. This is an important model to study in relation to the coal market, as it illustrates the maximum producer (or supplier) rent that can be generated. Cournot results lie somewhere between monopoly and perfect competition results.

Figure 6.2 below graphically represents a monopoly situation and the fact that the market price is no longer determined by the marginal cost. There is thus a premium charged to the consumers. The cost of monopoly is called dead weight loss. This dead weight loss is the reason why an economy should usually try to avoid monopolies.

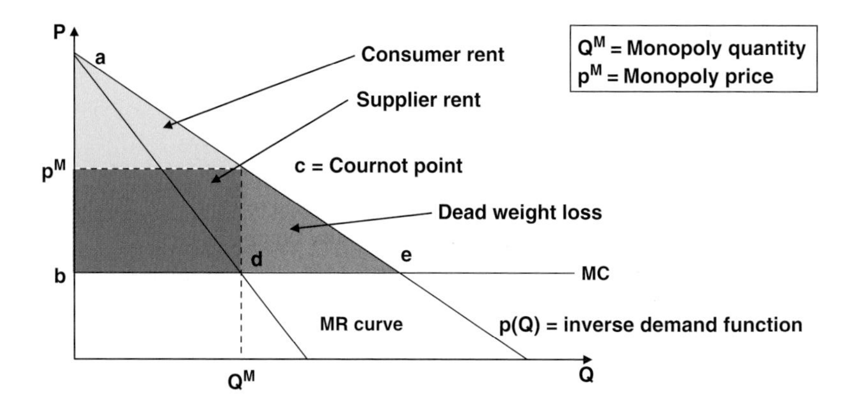

Fig. 6.2 Graphic representation of the reference model monopoly (Source: Author)

6.4 Cournot Competition

When studying Cournot, which means studying simultaneous quantity competition, one also should study Bertrand, meaning simultaneous price competition, and Stackelberg, meaning dynamic quantity competition. All three are relevant to the coal market:

- Cournot – simultaneous quantity competition: smaller players in the coal market take the market price as given, but here we study how they can retain some market power by adjusting their output quantity.
- Bertrand – simultaneous price competition: in some coal markets the output quantity may be given or fixed due to logistics or legislation; here I study how the market players have market power by adjusting their own price to the MC of the marginal producer for that market.
- Stackelberg – dynamic quantity competition: here I study how a new market entrant affects a monopolist's position. This is important for the coal market, as monopolists may be challenged by new entrants in smaller submarkets of the coal industry.

I will begin by looking at the simultaneous quantity competition, the classic Cournot model. As discussed previously, the Cournot model allows us to study competitive situations that are somewhere between perfect competition and monopoly. When looking for the Nash equilibrium in the Cournot game n symmetrical companies that compete on quantity are studied. Cournot differs from perfect competition as there are only n symmetrical firms and not an unlimited number of firms; it differs from monopoly as there is more than one firm. As such, it is somewhere in-between and in fact describes a market such as the coal market more realistically.

In Cournot, all companies act simultaneously and maximize their profit by adjusting their own supply quantity. As such, they have to react to every possible quantity of the other players; they therefore build a reaction function for it. Please see Appendix G for more details.

6.5 Bertrand and Stackelberg

Having looked at Cournot, simultaneous quantity competition with constant marginal cost in Section 6.4 and Appendix G, I now turn to Bertrand, simultaneous price competition, and Stackelberg, dynamic quantity competition.

6.5.1 Bertrand, Simultaneous Price Competition

In simultaneous price competition, the quantity is given and the market players compete on price for that given quantity. The quantity may be given or fixed due to logistics or legislation. The market power that each player has by adjusting his own price to the marginal cost of the marginal producer for that market can be illustrated.

When looking for the Nash equilibrium in the Bertrand game, n symmetrical companies compete on price and not quantity.

- Here it is assumed the consumers are extremely price-sensitive, much more so than before, and act accordingly.
- Unlimited production capacity is assumed: this means that any company can produce as much as it likes.
- Thus, anyone can flood the market and thus any player has a lot of market power (poses a threat to the other players).
- The Bertrand model determines the price, which in the end logically should equal the MC of the lowest cost producer; however, in a heterogeneous game the lowest MC producer is able to increase his price to the MC of the second lowest MC producer, and thus makes a profit.
- In this model different prices p_i are calculated.

In the Bertrand model companies act simultaneously and try to maximize their profits by adjusting their own supply price. As such, they have to react to every possible price of the other players; therefore, they build a reaction function for it. For more details please refer to Appendix G.

6.5.2 *Stackelberg, Dynamic Quantity Competition*

In the Cournot model the market players act simultaneously and compete on quantity. The Stackelberg model, on the other hand, examines dynamic quantity competition. Thus, here the players react to one another. The Nash equilibrium in a quantity game is still the goal. To analyze this case we first identify a Stackelberg Leader' who is already in the market and a 'Stackelberg Challenger' who is considering entering the market. The Leader anticipates the Challenger's move and reacts in order to maximize his profit.

From a coal market perspective, we may consider the following case: a monopolist (i.e., a coal producer that exclusively supplies a country or a consumer) is almost always attacked by potential new market entrants. A monopolist (the Leader) therefore tries to create barriers to entry, or, if that is not possible, anticipates a new market entrant (the Challenger) and sets his own new quantity in such a way that his profit is still maximized.

6.5.2.1 Conclusions of the Stackelberg Model

The Challenger chooses the quantity to offer in such a way that it equals the quantity that he would offer in the Cournot game. A first-mover (or early-mover) advantage exists because the Leader can set a higher output quantity and therefore has a higher profit than the Challenger. However, the Leader (initially the monopolist) has to give up part of his profits (as illustrated in formulas in Appendix G). Conclusions of the Stackelberg game with the assumptions set out in Appendix G:

- The market price is reduced from 9 to 6.
- The market quantity is increased from 6 to 9.
- The sum of supplier's profit is reduced from 36 to 27 because the Challenger's entry results in a higher quantity and therefore a lower price.
- The Leader's profit is reduced from 36 to 18.

6.6 Cournot with Constant vs. Increasing Marginal Cost

Having studied the classic Cournot, Bertrand, and Stackelberg game theory assuming constant marginal cost for all players, in Appendix G I went a step further and calculated how Cournot would change when increasing marginal cost is assumed. While this is not a novelty, the literature rarely speaks of the impact of increasing marginal cost on prices and profits. More important are the conclusions for corporate strategy in the coal business. The goal is to look at some intuitive market phenomena and compare them to Cournot results. All calculations have been made independent of prior research.

Most, if not all, natural resource markets are characterized by variable, increasing marginal cost. I have empirically demonstrated that this is the case for the coal market. Therefore, extending Cournot to study a game with increasing marginal cost is valuable for studying any natural resource market. My hypothesis is that the Cournot price with constant MC is lower than the price with increasing MC, all else being equal.

In this analysis I have relaxed one major assumption of the Cournot Game. I have calculated Cournot with increasing marginal cost. Below I compare the price, profit, and profit premium of cournot with constant and increasing marginal cost.

Cournot with constant MC	Cournot with increasing MC
$p^*(Q) = \dfrac{a - c_v}{n+1} + c_v$	$p^*(Q) = \dfrac{a(b + c_2) + bnc_1}{b(n+1) + c_2}$

Cournot with constant MC	Cournot with increasing MC
$\Pi_i^*(n) = \dfrac{1}{b}\left(\dfrac{a - c_v}{n+1}\right)^2 - C_f$	$\Pi_i^*(n) = \dfrac{(a - c_1)^2 b}{(bn + b + c_2)^2} - \text{FC}$

$$\Pi P = \frac{(a - c_1)^2 \left[-2bc_2(n+1) - c_2^2\right]}{(b(n+1) + c_2)^2\, b(n+1)^2} \xrightarrow{\lim c_2 \to \infty} \frac{-(a - c_1)^2}{b(n+1)^2}$$

Overall implications: I have shown that price increases with increasing marginal cost versus constant marginal cost and that profit decreases. It has been shown that in scarce resource markets where the MC curve becomes steeper prices increase, thus confirming the hypothesis and market intuition. More interestingly, it has been shown that firms should strive for a flat MC curve as their profits decrease, the steeper their MC curve becomes. Symmetrical firms had to be assumed, but in real life this becomes especially relevant for the marginal cost producers in a scarce market whose profits decrease, the steeper their MC curve is. However, firms that produce with costs below the cost of the marginal cost producer are very interested in scarce markets because the resulting price increase goes straight to their bottom line. As before, the market tends toward consolidation as fewer competitors also increase profits in markets with increasing marginal cost. For more details please refer Appendix G.

Chapter 7
Conclusions, Implications, and the Future of Coal

7.1 Introduction

In this book I have looked at the coal market from several angles: (1) I examined the source or supply side of steam coal exports including the major producing countries and supply concentration in Chapter 3; (2) I looked at the use of or demand for steam coal exports with particular focus on the power markets and environmental issues in Chapter 4; (3) I looked at the market as a whole in Chapter 5, based on primary research and a GAMS-programmed WorldCoal market model, focusing also on the geopolitical environment. In Chapter 6, I then discussed game theory, illustrated the well-known Cournot competition to include increasing marginal cost and the implications on price and profits.

In this chapter, I will make some predictions about the future while attempting to answer the questions raised. First, I will turn to the key implications of the Cournot extensions in Section 7.2. I will then look at the current and future market dynamics of the coal trade in Section 7.3 before making more general predictions about the price trends in Section 7.4 and the future of energy and coal in Section 7.5.

7.2 Implications of the Cournot Extension for the Coal Market

I expect the worldwide seaborne steam coal market to behave in a manner between the extremes of monopoly and perfect competition. Given the relative lack of transparency in the coal markets, the Cournot market model needs to be applied to the coal market in the same way as the perfect competitive model.

In their well-written paper and their CoalMod model Haftendorn and Holz (2008) have already explored how perfect competition and Cournot with constant marginal cost apply to the coal market for the years 2005 and 2006. They reached two key findings. First, for the years researched (2005 and 2006), they found that perfect competition best described the real price levels. This agrees with my conclusion based on the WorldCoal model built during fall 2007 (see Section 5.5 and Appendix E). Their second finding is that Cournot better describes the

L. Schernikau, *Economics of the International Coal Trade*,
DOI 10.1007/978-90-481-9240-3_7, © Springer Science+Business Media B.V. 2010

variation in reference prices between countries. Since Cournot allows for price discrimination and perfect competition does not, both models should be applied to the coal market.

In this study, Cournot's constant marginal cost assumption was relaxed to include increasing marginal cost and to discuss the impact on price and profits. Despite the simplifying assumptions discussed (Appendix G) that the competing companies are symmetrical and their cost function is quadratic, the following conclusions can be made about Cournot, confirming some market intuition:

1. Markets with increasing marginal cost, such as natural resource markets, demand a higher price premium under Cournot than markets with constant marginal cost, such as standard manufacturing-based markets.
2. A shortage situation in a natural resource market with steeper marginal cost curves results in a higher price premium in Cournot.
3. Symmetrical players with increasing marginal cost curves in a market (i.e., a scarce market) will earn less profit than players in a market with constant marginal cost curves (i.e., without scarcity).
4. The more players participate in an increasing marginal cost Cournot market, the lower the price and profits; the same applies to constant marginal cost Cournot markets.

For the coal market this means that a scarce market with steep marginal cost curves results in higher prices and lower profits for the marginal producer. On the other hand, non-marginal producers are interested in scarce markets as scarcity increases price and, hence, profits. As a real-world example let's take a supply market with a small number of fairly symmetrical coal exporters (e.g., Russia) and a demand market that is supplied almost exclusively by these exporters (e.g., Finland). Here one can well apply the new increasing marginal cost Cournot model. The exporters are able to translate their market power into increased profits. The Russians would not like a true scarce supply market; in fact, they would do everything possible to keep the market affluent and supply at the equilibrium price as described above. The scarcer the market becomes (or the steeper their marginal cost curve becomes), the less the profits they will be able to extract based on their market power.

From the above it can be concluded that market players have an incentive to continuously invest in their production to make the marginal cost curve 'flatter.' At the same time, investments will increase barriers to entry and thus play a part in protecting a player's market position. It has also been shown that all players in an increasing marginal cost Cournot market are interested in a reduced number of market participants; thus, consolidation will result. This has already been seen in the coal market, where the top five coal exporters in 2006 (BHP, Anglo, Xstrata/Glencore, Rio Tinto, and Drummond) control Australia, South Africa, and Colombia with about 67–86% of exports. Even in Russia and Indonesia these five players directly or indirectly account for almost 40% of exports (see Fig. 3.9)

I will focus more on supply concentration in the following section.

7.3 Current and Future Market Dynamics of the Coal Trade

The first transformation to global coal trade occurred in the new millennium at the beginning of this decade. Becker and Ungethuem (2001), formerly of Enron, have already seen a change in the market, where historically long-term contract buying by utilities changed to increased spot buying due to the need for pricing close to current market levels. Utilities used to buy from one or two mines and purchasing were driven by technical rather than market considerations. Today, there is more spot buying and we are in the midst of a trend in which a real global commodity market for coal has developed or is still developing.

7.3.1 Market Participants

Driven by utilities' need for market-driven pricing and new market participants, new and larger physical coal traders have developed. Traders increasingly handle financing and form a buffer between producers and consumers. They offer long-term off-takes to producers and offer spot-market-priced coal to consumers. As a result of this study and my research I now differentiate between four kinds of traders:

- vertically integrated, utility-led traders, such as EDF Trading, Total, RWE Trading, Constellation, E.ON Trading, Essent Trading, and others;
- vertically integrated, producer-led traders, such as Glencore, Noble, Vitol, and others;
- cash-rich independent physical volume traders very active in the Richards Bay and ARA volume markets, such as Flame, Bulktrading, Carbofer, Coal & Oil, and others; and
- medium-sized, specialized independent traders, such as Energy, Garcia Munte, HMS Bergbau, CC Carbon, and a number of others.

In the future, market participants will invest more and more in logistics and upstream assets. Coal remains a scarce raw material. This is supported by my conclusion (also by the IEA and others) that coal demand will continue to show higher growth rates than those for other traditional electricity generation methods such as nuclear, gas, and oil. I have shown that only 39% of global FOB costs in 2006 (see Fig. 3.10 on page 61) were pure mining costs. Thus, of CIF costs, we can conclude that pure mining accounts for only 10–20%, depending on freight. It is therefore evident that the coal market is largely about logistics and access to resources. This is apparent in the struggle for export port capacity, not only in South Africa but also in Russia, Australia, Colombia, and Indonesia.

7.3.2 Physical Trading Volumes

Steam coal is traded over large distances to reach consumers in Europe and Asia. Currently, less than 20% of globally consumed coal is traded between countries; the remainder is consumed in the country of production. In 2007, 782 million tons of hard coal was traded, of which 595 million tons was steam coal. Rickets (2007), from the IEA, mentioned that global hard coal trade volumes may reach 1.4 billion tons by 2020/2030. Assuming the historic steam coal trade growth of 5.4% per annum (see Section 5.2), I would extrapolate that the steam coal trade will reach 735 million tons in 2010 and 1.2 billion tons in 2020. These numbers are expected to be rather conservative, as even the VDKI expects 1 billion tons of trade by 2010. More recently, the trade growth rate has reached 8% per annum (World Coal Institute-Resource Coal 2005, p. 14). However, the financial crisis that started in fall 2008 will most certainly slow down growth for a year or two.

Globalization translates into increased trade. For coal this means that trading volumes will grow faster than the underlying growth in demand for coal. It can be seen in Fig. 7.5 that electricity demand from coal is expected to grow at a CAGR of 3.1% by 2030. The resulting even higher growth of the steam coal trade will further increase the significance of our market. These higher volumes will also professionalize the coal market further, attracting new, well-educated talent, and increase transparency.

7.3.3 Trading Risks

The growing importance of coal traders for financing and buffer purposes will also lead to more 'intra-trader' coal trading. Long trading chains of 10 or even more participants in one physical vessel delivery already developed during the price increases in 2007 and 2008. While price drops (in fall 2008 caused by the financial crisis and the overheating of the coal market) will reduce these trading chains temporarily, I expect that we will see each coal delivery go through more hands in the long term. As a result the risks for any trader increase, because default risk increases proportionally to the number of traders in one chain. These risks will also affect producers and consumers.

As price volatility increases, so does trading risk. As a result of the above developments, well-thought-out credit risk requirements and compliance are essential but probably not sufficient to keep trading as safe as it used to be. The development of physical and financial exchanges for coal will be sped up by these developments. For example, the physical OTC coal exchange globalCOAL represents one important step toward managing credit risks. But the market has to develop true exchanges that manage default risks themselves. The same is true for financial exchanges. The exchange-driven ICE platform already increased coal derivative volumes significantly during the months of the financial crisis in 2008. The same can be expected from the EEX in Germany or ASX in Australia. Real coal futures are emerging for both physical and paper coal. This development will accelerate in the years to come.

I predict that only the large standardized coal volumes will begin to be handled via exchanges in the next decade or so. This will include the standard RB-ARA route as well as NEWC-ARA. For the more fragmented trading routes that include 'off-spec' coal products it will become much more difficult to standardize. Hence,

I predict that at least 50% of the traded coal volume will continue being traded outside of exchanges, thus requiring physical coal traders to manage the product flow.

7.3.4 Financial Markets/Derivatives

Increased trade volumes will continue to attract new financial traders, including but not limited to investment banks. The recent financial crisis (this chapter was written in fall 2008) has temporarily curbed financial speculation and the influx of new market players. The crisis will certainly toughen credit risk requirements as well. However, once the global financial system returns to normal, the financial coal market will continue the growth path it started. Participants in the Online Coal Market Survey confirmed that price volatility will be one result of the increased paper market (see Appendix D). In terms of derivative volumes, it has already been shown in Section 5.3.2 that coal derivatives still have much room to grow to reach similar volumes to other commodity derivatives. The development of exchanges will further mitigate trading risks (see above) and give access to new and more market participants for physical coal and derivatives trading.

7.3.5 Regional Developments

The five main export regions – South Africa, Australia, Indonesia, Colombia/Venezuela, and Russia – will not benefit equally from the expected volume increase. Local demand in the various regions will curb export volume levels. I consider Australia and Russia to be most likely to increase export volumes. However, Russia is more at risk because the national strategy is to increase gas exports and use indigenous coal more within the country. Indonesia, relatively speaking, is expected to increase production most, but domestic demand will grow faster than production can grow. The fringe suppliers the United States and China will tend to supply more to the market when coal prices are high. Demand will be driven by the Pacific market, here mostly China and India. Thus, increased Russian volumes will be exported primarily from the Far East.

I expect that African coal exports – outside of South Africa – will develop and increase in the next two decades. Africa is very rich in coal deposits, localized in Mozambique, Zimbabwe, and Botswana in Eastern Africa, as well as in Western Africa. National and international agencies such as the BGR are not always fully aware of the true potential of reserves and resources, but new discoveries and restatements will rectify that situation soon. The one key problem that African supplier countries have is a lack of logistical infrastructure. Inland transportation systems and port infrastructure need to be built and wash plants need to be erected. Large coal consumers such as Vale have already decided to invest in African coal with an investment horizon of up to 5 years before the first significant volumes of coal can be exported. It will be interesting to see how quickly African coal will appear on the international seaborne steam coal market.

China is the wild card for global coal supply since it plays a special role as a fringe supplier. The country produces about 50% of global coal volumes and historically has been an important supplier to Japan, South Korea, and Taiwan. More recently, China has turned into a net importer, but uncertainty remains about future coal export volumes that may disrupt the global market. I do predict that, over a longer period of time, China will neither be a major net importer nor be a net exporter. However, short-term fluctuations will become more significant and will add to coal market volatility.

7.3.6 Production and Trade Consolidation

I discussed the results of the Cournot calculations in the previous chapter. Scientifically speaking, consolidation in the supply market (whether of traders or of producers) makes economic sense for shareholders. It can be seen in Fig. 7.1 that consolidation leads to higher EBITDA margins and higher total shareholder returns (TSR).

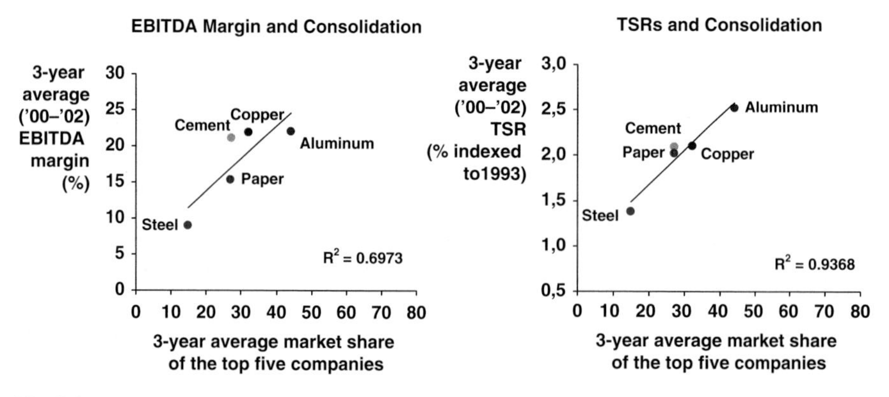

Fig. 7.1 Consolidation increases performance and returns to shareholders (Source: Woertler-BCG 2007)

From a macro-perspective, perfectly competitive markets are desirable as they reduce the so-called dead weight loss to zero. Thus, we have two trends, one driven by economics and one driven by politics. Policy will develop to better manage consolidation in the coal supply market, but it will not be able to stop consolidation. Historically, coal supply has not been regulated much. As far as policy and antitrust laws will allow, we will see more and more consolidation attempts, such as the merger talks between Rio Tinto and BHP. Also, the second-tier producers will continue to consolidate, either by merging with peers or by being bought up by larger players. The large drop in the valuation of raw material and coal producers that started in fall 2008 will, in fact, provide opportunities and could speed up consolidation.

7.4 Future Steam Coal Price Trends

The trend for future steam coal prices points toward higher prices. I come to this conclusion based on the research and analysis as part of this study. I have spoken to more than 100 industry participants as part of this work and much more than that as part of my daily working life. The reasons I have discovered are as follows:

- Electricity demand is going up;
- Coal's share of electricity generation is going up, mainly driven by China and India's large coal share in electricity generation;
- Major export countries, such as South Africa, Indonesia, and Russia, have increasing domestic coal demand;
- FOB costs will increase because of labor and transportation costs. Machinery and experienced personnel will likely remain in short supply;
- Export mine capacity utilization is increasing;
- Coal asset prices will increase as relative coal investments slowly catch up with oil and gas investments;
- Producers will continue to consolidate.

Figure 7.2 below depicts historic average steam coal for CIF ARA (API2) and FOB Richards Bay (API4). It can be seen that the first price increase occurred in

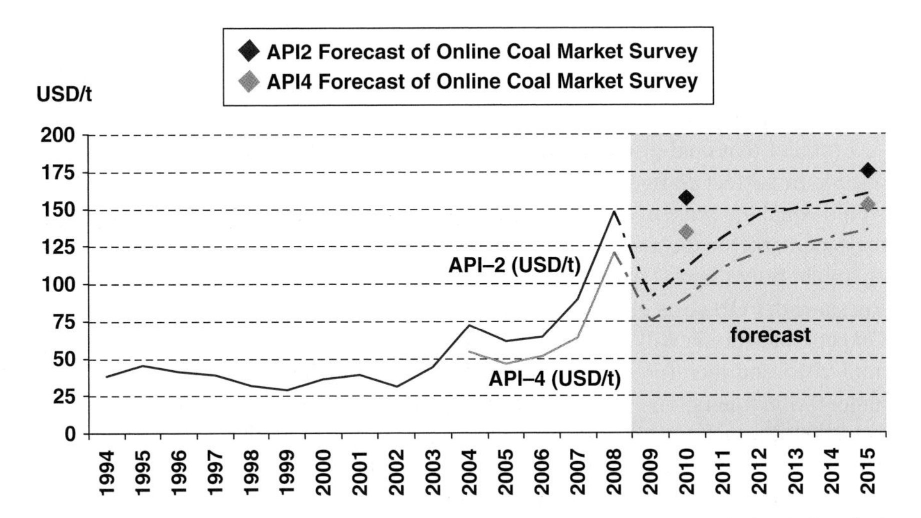

Fig. 7.2 Steam coal price trend until 2015 (Source: Author's research and analysis; Online Coal Market Survey; McCloskey Coal Price Index)

2003/2004. In 2005 and 2006 the market consolidated at a relatively high level before peaking in 2008. The average price for 2008 is tainted by the intra-year peak of US $210/ton for API2 in July. By comparison, the API2 was at US $130/ton in January 2008 and US $81/ton in December 2008.

I also attempt a coal price outlook until 2015. Admittedly, this outlook is very subjective. The graph above is meant to illustrate a trend that is driven by the fundamentals discussed previously and by the author's conclusion that future coal prices will tend to be above the marginal cost. Of course, it is impossible to foresee future coal prices, especially any intra-year volatility. As can be seen, I predict (today in fall 2008) that coal prices will soften in 2009, but still remain above 2007 levels, before slowly increasing again from 2010. The graph also shows the average coal prices predicted by the participants in our Online Coal Market Survey, which are above my personal predictions. However, the Online Coal Market Survey was conducted between August and October 2008, thus right in the middle of the financial crisis and the biggest coal price drop in history. From a separate analysis detailed in Appendix D it can be seen how individuals change their expectations about future prices in tandem with current coal price levels. I also predict coal price volatility to continue being very high, if not increasing further.

When looking at future coal prices, China is probably the single most important factor to consider. China may import as much as 700 million tons of steam coal by 2020. This would be about 100 million tons more than all of the 2006 traded steam coal volume. The current crisis will most definitely dampen China's growth rate for some time, but its long-term growth cannot be stopped. A more realistic scenario may be that China will increase domestic production in line with domestic coal demand and that China will not participate much in global coal trade.

Overall, I strongly believe that basic demand growth is driven by population growth, GDP per capita growth, and increasing electrification rates. Today, about 6.5 billion people inhabit our planet and by the middle of this century this number is expected to exceed 9 billion. At the same time, the average incomes of people in the so-called developing countries will continue to increase. The more money people have, the more they will spend on primary goods: water, food, and energy.

I predict that coal prices will rise above the marginal cost of the marginal producers. In perfect competitive markets, the marginal cost of the marginal producer would equal the price in equilibrium, but I conclude that this will not be the case for the coal market in the medium term. While I am not forecasting the FOB price curve or freight prices here, I expect that in 2008 the marginal cost producer will again be Russia with FOB costs between US $60 and 70/ton, depending on the exchange rate. I expect that Russia will also remain the marginal cost supplier to the world market until 2015 and therefore build a floor which – in my view – can only be undercut temporarily when global markets overreact to a crisis such as the one that began in fall 2008. Producers and traders have enough market power to keep average coal prices above perfectly competitive market price levels.

In relative terms, I expect coal prices to slowly catch up with gas prices. I believe that the basic economic principle of making a CO_2-friendly fuel less expensive than a non-CO_2-friendly fuel such as coal will win support from policymakers. I also believe that policymakers will eventually recognize that a European trading scheme is not the solution to the global CO_2 problem, but that it in fact contributes to it. I also expect that markets for sources of energy for the generation of electricity will become more co-integrated. I forecast that coal prices will be less volatile than, for instance, gas prices.

The price trend predictions leave out any expectation about future exchange rates. This paper has not considered exchange rates but they do remain an important driver for coal prices.

7.5 Future Source of Energy: What Role Will Coal Play in the Global Power Mix?

As can be seen in Fig. 7.3, global primary energy consumption is growing at an average annual rate (1990–2030) of about 1.7% (CAGR). Electricity consumption, on the other hand, has always grown much faster and will continue to do so at an average annual rate of 2.6% in the period 2005–2030. Growth is fueled by Non-OECD countries, most importantly China and India. As such, electricity generation will grow at a CAGR (2005–2030) of 4% in non-OECD countries versus 1.3% in OECD countries.

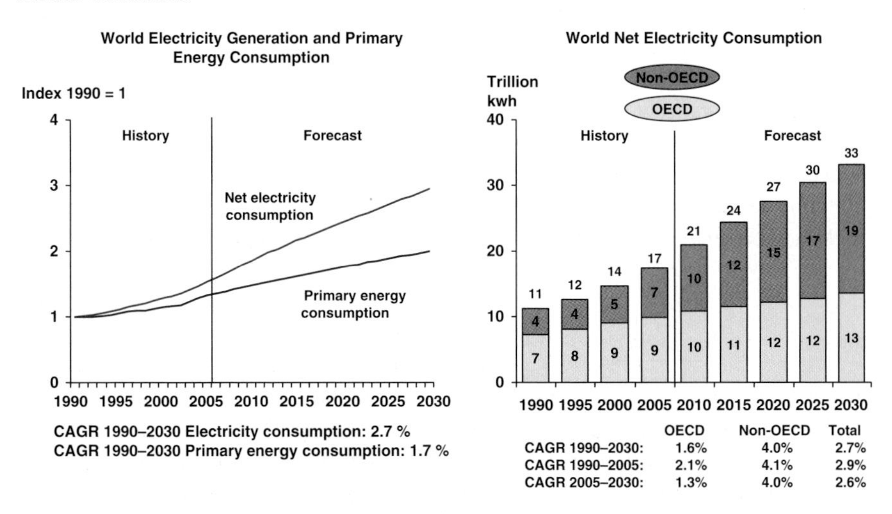

Fig. 7.3 Growth in world power generation and consumption, 1990–2030 (Source: Author's analysis; EIA 2008, Figs. 52 and 53)

Coal will continue to be the most important energy source for power generation and will increase its importance compared to oil and gas also for primary energy in general. Figure 7.4 illustrates how coal is expected to grow at a rate of 2.2% CAGR (2005–2030), faster than gas at 2.1% CAGR or even hydro at 2% CAGR or biomass at 1.3% CAGR. Only 'other renewables' are expected to grow faster at 6.7% CAGR. Interestingly, renewable energy sources as a whole can only keep their total 2005

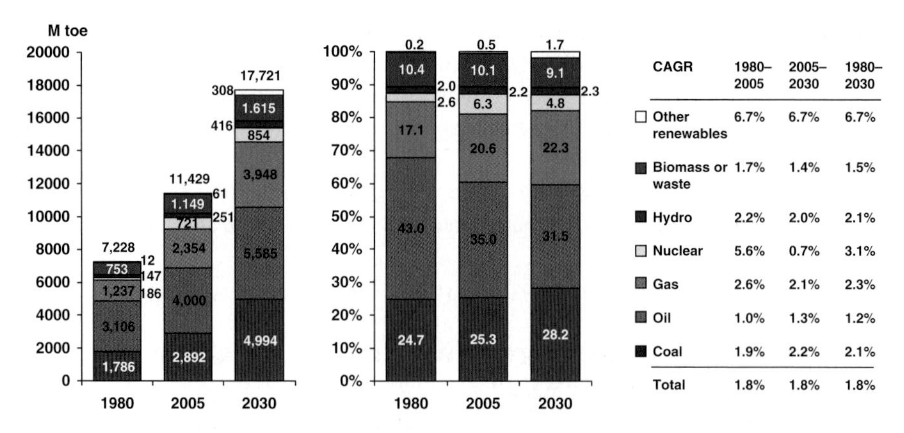

Fig. 7.4 IEA forecast of world primary energy consumption up to 2030 (Source: Author's analysis; IEA Energy Outlook 2007)

share of 13% of primary energy just about constant until 2030 (IEA Energy Outlook 2007).

It can be seen in Fig. 7.5 that coal is expected to increase its share in electricity generation to 46% by 2030. Thus, it is expected that almost half of all global electricity will be generated using coal by 2030. Only gas will continue to grow slightly more than coal with a CAGR (2005–2030) of 3.7 versus 3.1% for coal. As expected, oil's share will reduce. More interesting is the fact that renewable energy sources will retreat in relative terms. Renewable electricity generation simply cannot keep up with increased demand.

Future research could deepen our understanding of various aspects of the coal market. More extensive research could be applied to study, first, coal demand and,

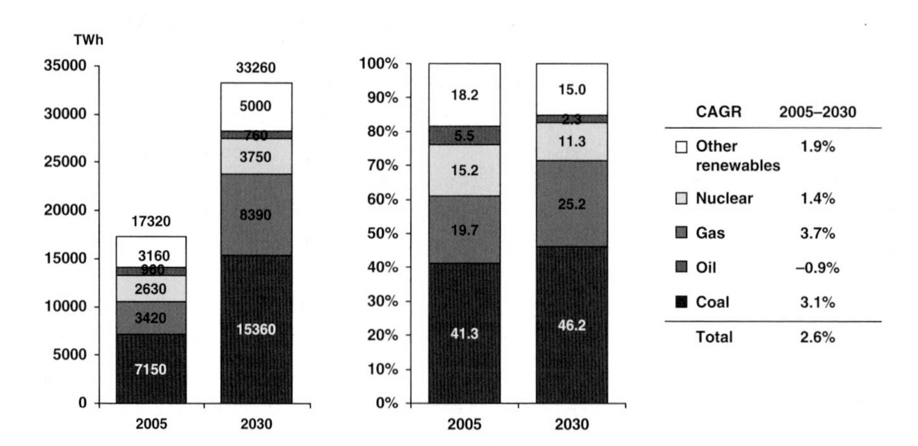

Fig. 7.5 Electricity growth by energy source until 2030 (Source: Author's analysis; IEA Energy Outlook 2007)

second, bulk freight. Thirdly, the analyzed coal supply regions can be broken down either to a company level or, better, to an export mine level. However, this would require major resources for traveling and interviews.

I believe study of the coal market could be usefully augmented by the following measures:

- The WorldCoal model could be updated to include newer data and detailed to include more supply and demand regions.
- Quality aspects, particularly heating values, of traded coal could be included in an updated global coal model.
- The investment cycles and costs involved with coal exploration and exploitation could be researched in more depth.
- Implications for the industry and equity markets could be analyzed.
- The new increasing marginal cost Cournot formulae can be applied to the coal market in a spatial equilibrium model and compared to perfect competition.
- The correlation between fuel prices for electricity generation, such as coal and gas, could be examined.

The impact of derivative markets, the development of exchanges for trading physical and derivative coal futures, and exchange rate fluctuations in raw material supply markets can also be researched in more detail as today's financial markets will increasingly influence the pricing and volatility of coal and other raw materials.

7.5.1 Environment

Environmental concerns about emissions of CO_2 from increased burning of coal will need to be addressed by the global community. I have argued that there is no way around coal as a source of electricity and energy in the coming decades. In fact, the use of coal will drastically increase. I predict that solar-based energy will only be able to reduce coal in the second half of this century. Politicians and scientists should, therefore, stop fighting coal and focus on improving technology to burn coal more efficiently. The best chances of reducing emissions lie in (a) more efficient power plants, (b) more efficient production of energy resources, (c) more efficient ways to transport energy resources and electricity, and (d) more efficient use of energy and electricity in general (UBS-Resources 2008, p. 29ff).

In order to support more gas-fired electricity (gas-fired generation produces about 45% less CO_2 per kW/h than coal-fired generation) it is important to close the gap between coal and gas prices. CO_2 trading schemes that penalize coal, which result in relative lower coal prices versus gas prices, in fact achieve the opposite. From a macroeconomic and geopolitical perspective, gas, the low-CO_2 fuel, should be priced below coal, the high-CO_2 fuel, and on a worldwide basis at that.

Independent of the current and future struggle to improve efficiencies, coal is likely to play an important role in replacing oil through CtL technologies. Coal may even play an important role in hydrogen production, a possible future source of energy. However, using coal as a source for hydrogen production will require successful carbon capture and storage technologies. CCS is a much-hoped-for technology to solve the big CO_2 problem of coal use. I personally believe that CCS will come but I do not think that it will be a long-term solution. In my view, the only long-term solution to all environmental problems associated with the burning of fossil fuels is solar energy.

The planet has access to enough energy in the form of solar radiation. The task for the future is not to develop new ways to release energy, but to discover how the primary needs of humankind (food, heat, electricity, and fuel/process energy) can be met with regenerative or renewable sources of energy. Although this study is about coal, I would like to conclude by repeating the formula proposed by Wolf and Scheer (2005) in their book *Öl aus Sonne – Die Brennstoffformel der Erde* (Oil from the Sun – The Earth's Fuel Formula), which illustrates that only solar energy is required to process carbon dioxide and water into a carbon-based fuel and oxygen:

$$\text{solar energy} + CO_2 + H_2O \Leftrightarrow CH_2 + 1.5O_2.$$

7.6 Suggestions for Future Research

Outside of this book, the economics of global coal trade have not been widely studied, especially when compared to gas, oil, or renewable energy sources. Therefore, there remains an open field for studying the coal market in more detail. I view this book as comprehensive introduction to coal with the goal of making the reader aware of the importance of steam coal in today's market.

Bibliography

Adelman, Morris A. and G. Campbell Watkins (2008). Reserve Prices and Mineral Resource Theory, The Energy Journal, IEAA, Special Edition 2008, 16, s162.

Argus Coal Daily International. Daily Coal Industry Publication, published by Argus Group Ltd., Brookfield, WI, when this sources is cited, the date of the report will be noted, s143.

Argus/McCloskey Index Report. Argus/McCloskey's Coal Price Index Report, Argus Media Group and McCloskey Group, issue 286, 27 June 2008 or Stated Separately, s195.

Baker, Kenneth (2000). Gaining Insight Into Linear Programming from Patterns in Optimal Solutions, Hannover, New Hampshire, 2000, online under http://archive.ite.journal.informs.org/ Vol1No1/Baker/, accessed 23.07.08, s196.

Baruya, Paul (2007). Supply Costs for Internationally Traded Coals, EIA Clean Coal Centre, London, July 2007, s161.

BCG (2004). Entwicklung eines Marktmodels für den Kohlemarkt, The Boston Consulting Group Germany, Munich, Germany, September 2004, s049a.

Becker, Sven and Manfred Ungethuem (2001). Der Kohlenmarkt im Wandel, Wirtschaftswelt: Energy, Sven Becker and Manfred Ungethuem, Enron, May 2001, s073.

BGR (2006). Reserven, Ressourcen und Verfügbarkeit von Energierohstoffen 2006, Bundesanstalt für Geowissenschaften und Rohstoffe (Federal Institute for Geosciences and Natural Resources), 23 November, 2007, s114.

BGR-China (2007). Commodity Top News No. 27: Die Rolle Chinas auf dem Weltsteinkohlemarkt, Sandro Schmidt, BGR Bundesanstalt für Geowissenschaften und Rohstoffe, Hannover, 8 January, 2007, s093.

BP Statistical Review (2006). Quantifying Energy – BP Statistical Review of World Energy June 2006, London, UK, June 2006, s038.

Brown, Stephen P.A. and Mine K. Yücel (2008). What Drives Natural Gas Prices? The Energy Journal, 29(2), 45, IAEE, s163.

Buisson, Nadine (2007). Industrie Mondiale Du Carbon, Evolution des Acteurs Producteurs/ Exportateurs, Nadine Buisson, 31 May 07 "Seminar La Dynamique des Marches Charbonniers", Dauphine University, Total, s112.

Cayrade, Patrick (2004). Investments in Gas Pipelines and Liquefied Natural Gas Infrastructure. What is the Impact on the Security of Supply? Fondazione Eni Enrico Mattei. Nota di Lavoro 114.2004 (http://www.feem.it/Feem/Pub/Publications/Wpapers/default.htm), s186.

China Coal Monthly (2006). China Cracks Down on Coal-to-Oil Frenzy, McCloskey's ChinaCoalMonthly, Issue 31, August 2006, s072.

China Daily (2004). Coal Mining: Most Deadly Job in China, by Zhao Xiaohui & Jiang Xueli (Xinhua), http://www.chinadaily.com.cn/english/doc/2004-11/13/content_; 391242.htm, updated: 13 November 2004, accessed: 4 June, 2008, s172.

Clarksons (2006). Freight: Smooth Sailing or Rough Seas? Henriette van Niekerk, Clarkson, Presentation at Coaltrans Athens, Greece, October 2006, s146.

L. Schernikau, *Economics of the International Coal Trade*,
DOI 10.1007/978-90-481-9240-3,

Cunningham, Eoghan (2007). Does ARA Need Physical Delivered Futures? CEO Globalcoal, Presentation at Coaltrans Rome, Italy, October 2007, s191.

Dantzig, George B. (1963). Linear Programming and Extensions, Princeton University Press, Princeton NJ, s197.

Davidson, Robert (1994). Abstract of Nitrogen in Coal, IEA Paper, published in January 1994, http://www.caer.uky.edu/iea/ieaper08.shtml, Read 31 March, 2008, s159.

Deutsche Bank Research (2007).; Technology to Clean up Coal for the Post-Oil Era, Deutsche Bank AG, Frankfurt am Main, 26 February, 2007, s097.

Deutsche UFG (2006). Suek Valuation Performed by Deutsche Bank based on 2005 Financial Data, published in 2006, s055.

Dlamini, Kuseni (2007). Richards Bay Coal Terminal: Contributing to SA's Growth and Development, South African Coal Conference, Presentation by Executive Chairman of RBCT, Capetown, South Affrica, 30 January, 2007, s140.

EIA (2007). EIA Annual Energy Review 2006; Energy Information Agency, Washington, DC, June 2007, s154.

EIA (2008). International Energy Outlook 2008, Energy Information Agency, Washington, DC, USA, September 2008, http://www.eia.doe.gov/oiaf/ieo/index.html, s209.

Einstein, Albert (1955). US (German-born) physicist (1879–1955), Quotations Page, http://www.quotationspage.com/quote/9.html, accessed 18 February, 2009, s219.

Endres, Stefan (2008). Eine Analyze des Weltmarkts für Steinkohle – Das Kohlemarktmodell WorldCoal, Technical University, TU-Berlin Energy Systems, Berlin, July 2008, s180.

Erdmann, Georg (2007). Gut Gemeint – das Gegenteil von Gut Gemacht: Das Kyoto Protokoll und die Globalen CO_2-Emissionen, erschienen in Energiewirtschaftlichen Tagesfragen 1/2 2007, S. 92–94, s176.

Erdmann, Georg and Peter Zweifel (2008). Energieökonomik: Theorie und Anwendungen, Springer-Verlag, Berlin; Heidelberg, s098.

Ernst & Young (2006). Energiemix 2020: Szenarien für den deutschen Stromerzeugungsmarkt bis zum Jahr 2020, Dr. Helmut Edelmann, Ernst & Young AG, Deutschland, s094.

European Commission (2005). Annual Report on the Implementation of the Gas and Electricity Internal Market; Report from the Commission, Brussels, Belgium, 5 January, 2005, COM(2004) 863 final, s174.

European Commission (2006). World Energy Technology Outlook – 2050: WETO-H_2, Brussels, Belgium, s136.

Faizoullina, Tatiana (2006). Coal Mine Methane Projects under Kyoto Protocol, EcoSecurities Group PLC; Tatiana Faizoullina, Athens, Greece, 27 September, 2005, s085.

Financial Times Deutschland (2006). Nato warnt vor Russischem Gaskartell, Financial Times Deutschland, 14 November, 2006, Article written by Schmid, Dombey, Steinmann, Zapf, s68.

Fisher, Franklin M. (1961). The Stability of the Cournot Oligopoly Solution: The Effects of Speeds of Adjustment and Increasing Marginal Costs; The Review of Economic Studies, 2(2), 125–135, February, s220.

Frachtkontor Junge (2007). Notizen zum Trockenmarkt, Frachtkontor Junge & Co., Hamburg, 23 March, 2007, s084b.

Frondel, Manuel, Rainer Kambeck and Christoph M. Schmidt (2007). Hard Coal Subsidies: A Never Ending Story? Rheinisch-Westfälisches Institut für Wirtschaftsforschung, RWI, Essen, Germany 16 Januar,y 2007, s153.

Galbraith's (2008). Galbraith's Dry Bulk Review, January 2008, s132.

Global Insight-Russia (2007). Global Insight: Analysis of the Russian Coal Market 2007, Global Insight, Inc., Boston, MA, s168.

Global Insight-World (2007). Global Insight: Global Steam Coal Trade & Price Forecast (2007-2035), pre-release from Global Insight, Inc., Boston, MA, s167.

Große, Mario, Lars Wischhaus, Robert Reck and Robert Wand (2005). Struktur und Preisbildung auf Steinkohlemärkten, TU-Dresden Energy Economics, December 2005, s056.

Haftendorn, Clemens and Franziska Holz (2008). Analysis of the World Market for Steam Coal, Using a Complementarity Model, Institute for Economic Research (DIW), Berlin, 5 August, 2008, s179.

Hartley, Peter R., Kenneth B. Medloc and Jennifer E. Rosthal (2008). The Relationship of Natural Gas to Oil Prices; International Association for Energy Economics, Energy Journal, 29(3), s177.

Holz, Franziska, Christian von Hirschhausen and Claudia Kemfert (2008). A Strategic Model of European Gas Supply (GASMOD), Energy Economics, 30(3), 766–788, Online under: http://www.tu-dresden.de/wwbwleeg/publications/ publications.html, accessed 23 July, 2008, s198.

Hotelling, Harold (1931). The Economics of Exhaustible Resources, Journal of Political Economy, 39, s162a.

IEA-CO_2 (2006). CO_2 Emissions from Fuel Combustion, 2006 Edition, International Energy Agency, Paris, s155.

IEA-Manual (2006). IEA Handbuch Energiestatistiken, IEA Publications, 9 rue de la Fédération, 75739 Paris Cedex 15, Printed In France By Stedi, January 2006, s039.

IEA-Oil (2007). IEA Statistics – Oil Information 2007, International Energy Agency, Paris, s206.

IEA-Statistics (2005). Key World Energy Statistics 2005, International Energy Agency, Paris, s034.

IEA CIAB-Coal Industry Advisory Board (2005). Reducing Greenhouse Gas Emissions – The Potential of Coal, IEA Publications, International Energy Agency, Paris, April 2005, s040.

IEA CIAB CtL (2006). CtL Workshop Results, IEA Coal Industry Advisory Board Workshop, IEA Headquarters in Paris, France, 2 November 2006, s189.

IEA Clean Coal Center (2008). Web Resource http://www.coalonline.org/catalogues/coalonline/81591/6247/html/6247_27.html, accessed 31 March, 2008, s158.

IEA Energy Outlook (2007). World Energy Outlook 2007: China and India Insights, International Energy Agency, Paris, OECD/IEA 2007, s123.

IEA International Energy Outlook-Electricity (2007). International Energy Outlook, Chapter 6: Electricity, 61–71, International Energy Association, Paris, s187.

IEA World Energy Investment Outlook (2003). World Energy Investment Outlook 2003, International Energy Agency, Paris, OECD/IEA 2003, s169.

Indonesian Mining Association (2006). Unternehmen Müssen in Kohlenbergbau Investieren, Indonesian Mining Association Press Article, October 2006, http://www.ima-api.com, s070.

Joskow, Paul (1987). Contract Duration and Relationship-Specific Investments: Empirical Evidence from Coal Markets, American Economic Association, American Economic Review, 77, 168–85, s060.

Kjaerstad, Jan and Filip Johnsson (2008). The Global Coal Market: Future Supply Outlook – Implications for the European Energy System, Department of Energy Conversions, Chalmers University of Technology, Gothenburg, Sweden, Pre-release paper, s175.

Kolstad, Charles D. and David S. Abbey (1983). The Effect of Market Conduct on International Steam Coal Trade, Los Alamos, USA, August 1983, European Economic Review 24 (1984), s181.

Kopal, Christoph (2007). Entwicklung und Perspektiven von Angebot und Nachfrage am Steinkohlenweltmarkt, ZfE Zeitschrift für Energiewirtschaft, 2007 no. 1, s157.

Kopel, Michael (1996). Simple and Complex Adjustment Dynamics in Cournot Duopoly Models, Department of Managerial Economics and Industrial Organization – University of Technology Vienna, Chaos, Solutions and Fractals, 7(12), 2031–2048, s221.

Krüger, Anke (2007). Neue Kohlenstoffmaterilien. Eine Einführung, Teubner, Abschnitt 1.3.2, s214.

Latham, Michael (2008). Efficient Mine Trucking Operations, Presentation by Michael Latham, Caterpillar Indonesia, Bali Coal Conference, Bali, Indonesia, 3 June, 2008, Stream 3 “Mining Operations”, s170.

Li, Raymond (2008). International Steam Coal Market Integration, Macquarie University, Department of Economics, Sydney, s182.

McCarl, Bruce A. and Thomas H. Spreen (2008). Applied Mathematical Programming Using Algebraic Systems, Kap. 5, S. 1–43, Online under http://agecon2.tamu.edu/people/faculty/mccarl-bruce/regbook.htm, accessed 23 July, 2008, s199.
McCloskey (2007). McCloskey Publishes Weekly Coal Reports, When This Sources Is Referenced, the Number and Date of the Report Will Be Noted, s147b.
Mimuroto, Yoshumitsu (2000). An Analysis of Steaming Coal Price Trends – Factors Behind Price Fluctuations and Outlook, Coal Research Group, International Cooperation Department, s194.
PESD Stanford (2009). Global Coal Market Conference, 2009 PESD Annual Winter Working Seminar, Program on Energy and Sustainable Development, Stanford University, Stanford, CA, February 2009, s215.
Petromedia Ltd. (2008). Portworld – Distance Calculator, Online under: http://www.portworld.com/map/, accessed 23 July, 2008, s200.
Pinchin-Lloyd's (2005). Lloyd's Maritime Intelligence Unit, Lloyd's List – Ports of the World, London, UK, s201.
Rai, Varun, David G. Victor and Mark C. Thurber (2008). Carbon Capture and Storage at Scale: Lessons from the Growth of Analogous Energy Technologies, Program on Energy and Sustainable Development (PESD), Stanford University, Stanford, CA, November, s216.
RBCT (2008). Richards Bay Coal Terminal Website, www.rbct.co.za, accessed 2 April, 2008, s160.
Rickets, Brian (2007). A Review of Steam Coal Supply/Demand Within IEA World Energy Outlook, IEA, Coaltrans Rome, Italy, October 2007, s193.
Ritschel, Wolfgang and Hans-Wilhelm Schiffer (2007). World Market for Hard Coal, 2007 Edition, RWE Power, Essen/Cologne, October 2007, s152.
Rosenthal Richard. E. (2008). *GAMS* – A User's Guide, Tutorial, S. 5-25, GAMS Development Corporation, Washington, DC, s203.
Schagermann, Chris (2007). Liquidity in the Coal Market, Energy Trading BHP Billiton, Coaltrans Rome, Italy, October, s190.
Schwarz, Sarah (2006). Thesis: Untersuchung der Korrelation von Importsteinkohlen- und Strommarkt als Voraussetzung für die Bewirtschaftung von Strom- und Brennstoffportfolios, Technische Universität Braunschweig; March; s066.
Simes, Bill and Barlow Jonker (2006). Thermal Coal Market Outlook, Bill Simes, Executive Director Barlow Jonker, Athens, Greece, October 6, s110.
Stanford (2008). The Political Economy of the Global Coal Market, Program on Energy & Sustainable Development (PESD), Stanford, CA, February, s213.
Steenblik, Ronald P. and Panos Coronyannakis (1995). Reform of Coal Policies in Western and Central Europe: Implications for the Environment, Energy Policy 23(6), 537–553, s171.
Tester, Jefferson (2009). A Pathway for Widespread Utilization of Geothermal Energy, Presentation at Stanford University by Professor Jefferson Tester from MIT, Stanford, CA, February, 2009, s217.
Tirole, Jean (1988). The Theory of Industrial Organization. MIT Press, Cambridge, MA, s185.
Tory, Richard (2006). Lehman Brothers: Global Financial Market Appetite for Investment in Coal, Presentation at Coaltrans World Coal Conference, Athens, Greece, 23 October, 2006, s108.
UBS-Resources (2008). UBS Research Focus – Knappe Ressourcen als Herausforderung und Chance, Wellershoff and Reiman; UBS AG Wealth Management Research, Zurich, August 2008, s207.
Uni Basel (2007). Preiseleastizität, R.H. Frank: Microeconomics and Behavior, McGraw Hill Inc., New York, NY, http://www.wwz.unibas.ch/wi/projects/toolkit/lektionen/lpreis.html, accessed 2 December, 2007, s208.
Vahlenkamp, Thomas and James O. McKinsey (2006). Integrated Power Perspective – Implications for Regulation, Thomas Vahlenkamp, McKinsey, Coaltrans Athens, Greece, October, 2006, s106.
Varian, Hall R. (1999). Grundzüge der Mikroökonomik, 4. Aufl., pp. 274–294, Munich, Vienna and Oldenbourg, 1999, s202.

VDEW (2007). Lexikon der Energiewelten, VDEW, http://www.energiewelten.de/elexikon/lexikon/index3.htm, accessed 26 December, 2007, s173.
VDKI Annual Report (2006). Annual Report of the German Coal Importers Association (Verein der Kohlenimporteure e.V.), Hamburg, Germany, Spring 2007, s011c
VDKI Annual Report (2007). Annual Report of the German Coal Importers Association (Verein der Kohlenimporteure e.V.), Hamburg, Germany, July 2008, s011d
VGB (2003). Konzeptstudie, Referenzkraftwerk Nordrhein-Westfalen (RKW NRW), VGB PowerTech e.V., Essen, Germany, 19 November, 2003, s188.
Warell, Linda (2007). Market Integration in the International Coal Industry: A Cointegration Approach, Lulea University of Technology, Economics Unit, Luleå, 2006, 38, s183.
Weaving, Derek (2006). Reform of Russian Power and Gas, Derek Weaving & Co., McCloskey's 4th Annual Russian Coal Conference, St. Petersburg, Russia, 2006, s103.
Woertler, Martin-BCG (2007). Limitless Resources to Fuel the Growth "Five Hypotheses on Raw Materials", BCG 1st European Alumni Meeting, The Boston Consulting Group Industrial Goods, Martin Woertler, Munich, Germany, 26 November, 2007, s139.
Wolf, Bodo and Herrmann Scheer (2005). Öl aus Sonne – Die Brennstoffformel der Erde, Bochum, Germany, 2005, s001.
World Coal Institute-Resource Coal (2005). The Coal Resource: A Comprehensive Overview of Coal, World Coal Institute, London, First published in the UK, May 2005, s041.
World Coal Institute-Secure Energy (2005). Coal: Secure Energy, World Coal Institute, London, First published in the UK, October 2005, s042.
World Wide Fund for Nature (2007). Dirty Thirty – Ranking of the Most Polluting Power Stations in Europe, World Wide Fund for Nature, Brussels, Belgium, May, 2007, s131.
Yaxley, Nigel (2006). President Euracoal, Secure and Sustainable Energy from Coal, Presentation in Brussels, Germany, 23 January, 2006, s014.

Coal seam in South Africa

Thirty-meter low-rank coal seam in Kalimantan, Indonesia

Small KP coal mining in Indonesia

Coal outcrop in Kalimantan, Indonesia

Port of Maputo loading operation

Reclaimer in Southern Africa

Tippler operation in Mozambique Port

Barge loading operation in South Kalimantan, Indonesia

Appendix A
Conversion of Energy Units

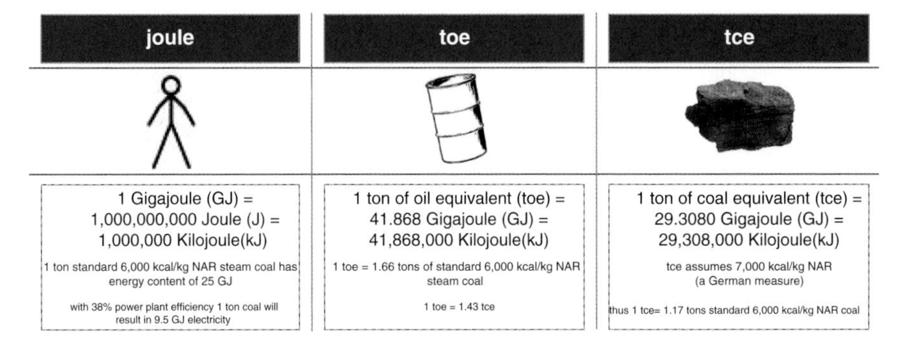

Definition of energy units

Input \ Output	kJ	kcal	kWh	Kg ce	kg oe	m^3 natural gas
1 kJ	1	0.2388	0.000278	0.000034	0.000024	0.000032
1 kcal	4.1868	1	0.001163	0.000143	0.0001	0.00013
1 kWh	3,600	860	1	0.123	0.086	0.113
1 kg ce	29,308	7,000	8.14	1	0.70	0.924
1 kg oe	41,868	10,000	11.63	1.428	1	1.319
1 m^3 natural gas	31,736	7,580	8.816	1.082	0.758	1

Conversion of energy units

Appendix B
World Electricity Production by Fuel

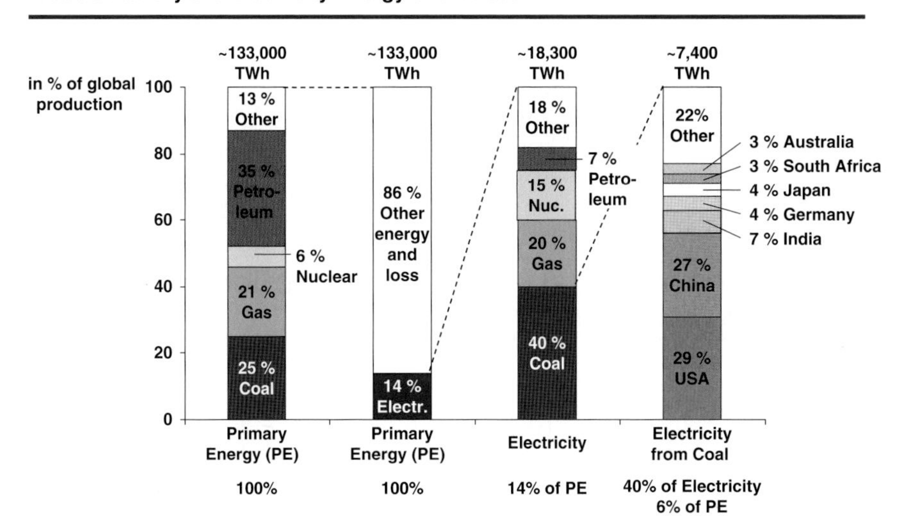

COAL: 40% OF GLOBAL ELECTRICITY PRODUCTION (I)

World Electricity Energy Share 2005

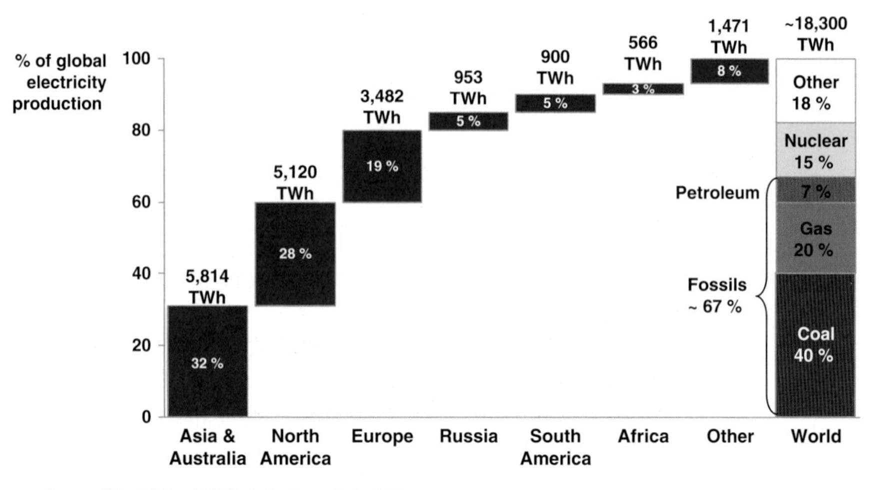

Source: IEA statistics (2005); Author's analysis, i019

COAL: 40% OF GLOBAL ELECTRICITY PRODUCTION (II)

World's Electricity Energy Share 2005

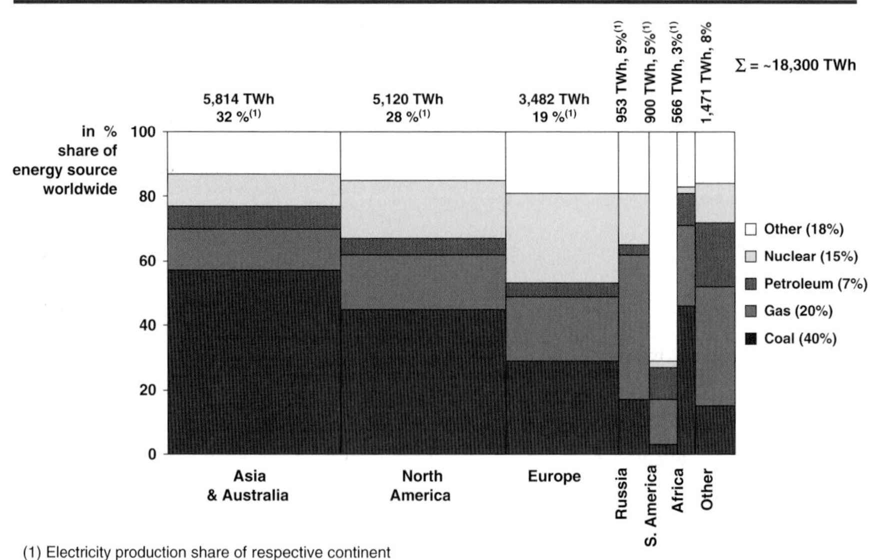

(1) Electricity production share of respective continent
Source: IEA statistics 2005; Author's analysis, i203

TOP 10 ELECTRICITY PRODUCERS...
...account for 64% of Global Electricity Production

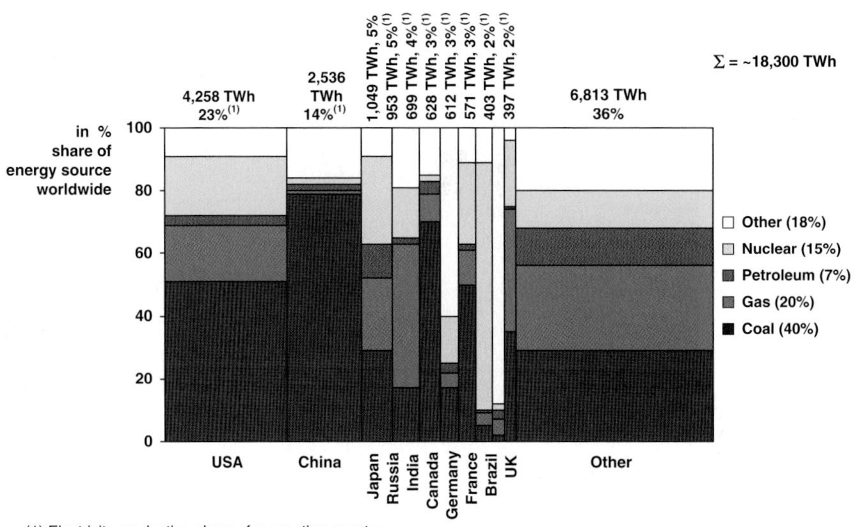

(1) Electricity production share of respective country
Source: IEA statistics 2005; Author's analysis, i020

TOP 10 COAL CONSUMERS FOR ELECTRICITY...
...account for 83% of Coal Consumption for Electricity

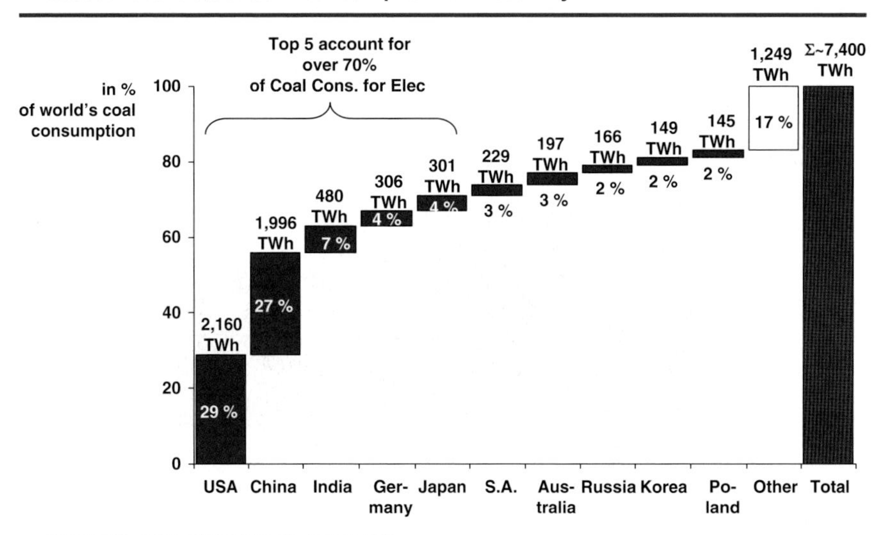

Source: IEA statistics 2005; Author's analysis, i205

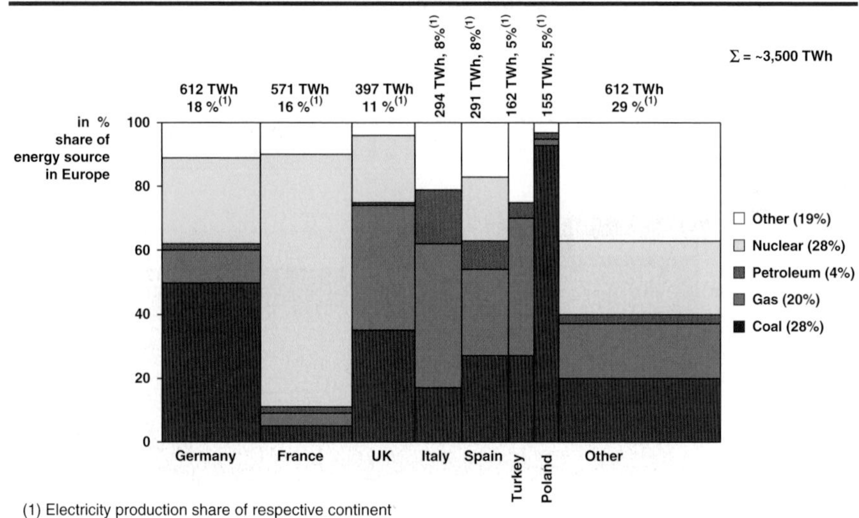
COAL: 28% OF EUROPE'S ELECTRICITY PRODUCTION
European Countries' Electricity Energy Share 2005
Σ = ~3,500 TWh
612 TWh 18 %(1)
571 TWh 16 %(1)
397 TWh 11 %(1)
294 TWh, 8%(1)
291 TWh, 8%(1)
162 TWh, 5%(1)
155 TWh, 5%(1)
612 TWh 29 %(1)
in % share of energy source in Europe
100
80
60
40
20
0
Germany
France
UK
Italy
Spain
Turkey
Poland
Other
Other (19%)
Nuclear (28%)
Petroleum (4%)
Gas (20%)
Coal (28%)
(1) Electricity production share of respective continent
Source: IEA statistics 2005; Author's analysis, i207

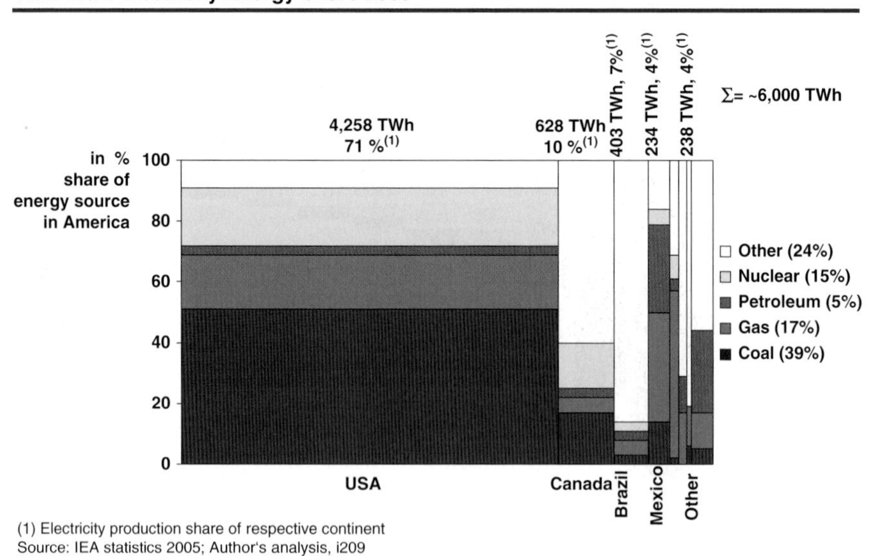
COAL: 39% OF AMERICA'S ELECTRICITY PRODUCTION
America's Electricity Energy Share 2005
Σ= ~6,000 TWh
4,258 TWh 71 %(1)
628 TWh 10 %(1)
403 TWh, 7%(1)
234 TWh, 4%(1)
238 TWh, 4%(1)
in % share of energy source in America
100
80
60
40
20
0
USA
Canada
Brazil
Mexico
Other
Other (24%)
Nuclear (15%)
Petroleum (5%)
Gas (17%)
Coal (39%)
(1) Electricity production share of respective continent
Source: IEA statistics 2005; Author's analysis, i209

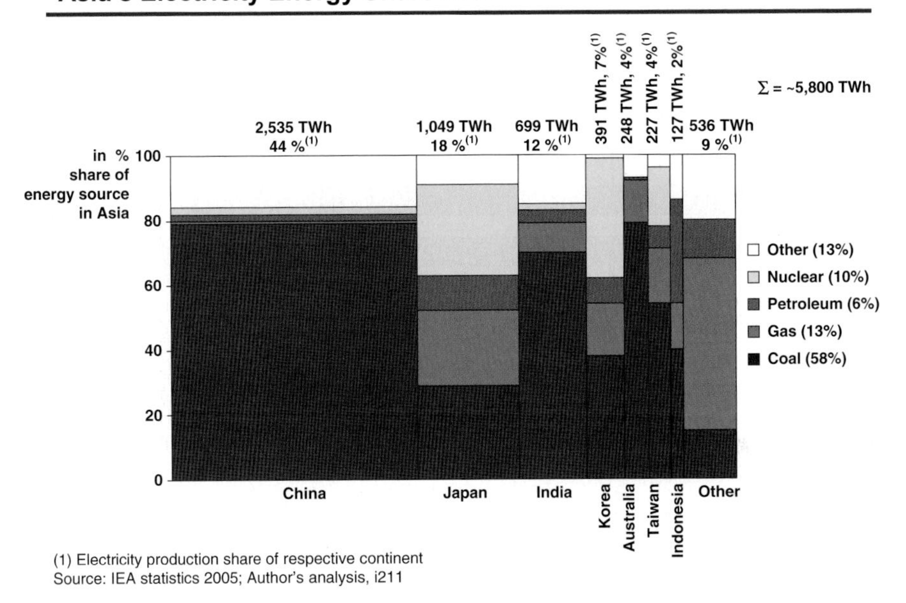
COAL: 58% OF ASIA'S ELECTRICITY PRODUCTION
Asia's Electricity Energy Share 2005
Σ = ~5,800 TWh
2,535 TWh 44 %(1)
1,049 TWh 18 %(1)
699 TWh 12 %(1)
391 TWh, 7%(1)
248 TWh, 4%(1)
227 TWh, 4%(1)
127 TWh, 2%(1)
536 TWh 9 %(1)
in % share of energy source in Asia
100
80
60
40
20
0
China
Japan
India
Korea
Australia
Taiwan
Indonesia
Other
Other (13%)
Nuclear (10%)
Petroleum (6%)
Gas (13%)
Coal (58%)
(1) Electricity production share of respective continent
Source: IEA statistics 2005; Author's analysis, i211

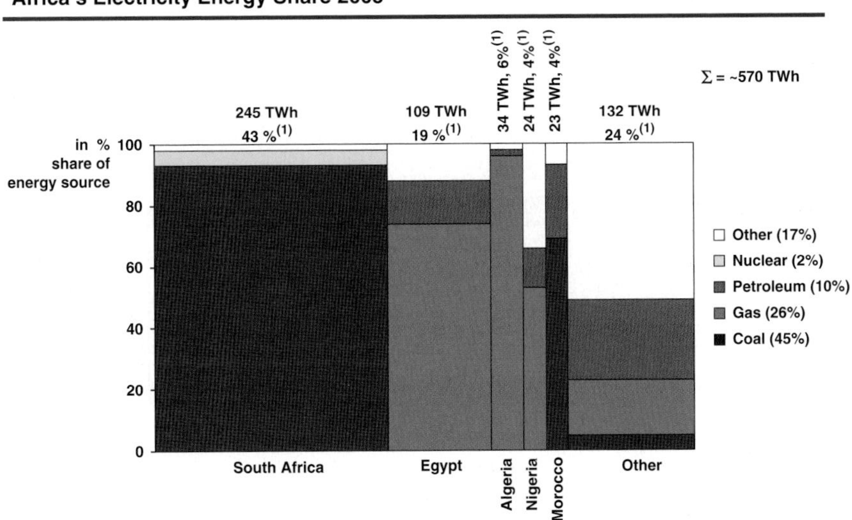
COAL: 45% OF AFRICA'S ELECTRICITY PRODUCTION
Africa's Electricity Energy Share 2005
Σ = ~570 TWh
245 TWh 43 %(1)
109 TWh 19 %(1)
34 TWh, 6%(1)
24 TWh, 4%(1)
23 TWh, 4%(1)
132 TWh 24 %(1)
in % share of energy source
100
80
60
40
20
0
South Africa
Egypt
Algeria
Nigeria
Morocco
Other
Other (17%)
Nuclear (2%)
Petroleum (10%)
Gas (26%)
Coal (45%)
(1) Electricity production share of respective continent
Source: IEA statistics 2005; Author's analysis, i213

WORLD'S ELECTRICITY ENERGY SHARE 2005

Source	World		Asia & Australia		North America		Europe (OECD)		Russia	
	[TWh]	%	[TWh]	%	[TWh]	%	[TWh]	%	[TWh]	%
Coal	7,378	40	3,377	58	2,297	45	991	28	166	17
Natural Gas	3,606	20	760	13	893	17	700	20	439	46
Petroleum	1,208	7	368	6	232	5	141	4	21	2
Nuclear	2,783	15	553	10	912	18	980	28	149	16
Other	3,332	18	756	13	787	15	670	19	178	19
Total	18,307	100	5,814	100	5,120	100	3,482	100	953	100

Source	South America		Africa		Other	
	[TWh]	%	[TWh]	%	[TWh]	%
Coal	28	3	251	45	268	18
Natural Gas	124	14	149	26	542	37
Petroleum	95	11	59	10	292	20
Nuclear	17	2	11	2	160	11
Other	636	70	96	17	209	14
Total	900	100	566	100	1,471	100

World Share	
Asia	32%
North America	28%
Europe	19%
Russia	5%
South America	5%
Africa	3%
Other	8%

Source: IEA statistics 2005; Author's analysis, i214

TOP TEN ELECTRICITY PRODUCING COUNTRIES 2005

Source	USA		China		Japan		Russia		India		Canada	
	[TWh]	%	[TWh]	%	[TWh]	%	[TWh]	%	[TWh]	%	[TWh]	%
Coal	2,160	51	1,996	79	301	29	166	17	480	70	104	17
Natural Gas	776	18	26	1	239	23	439	46	62	9	32	5
Petroleum	143	3	61	2	122	11	21	2	31	4	21	3
Nuclear	810	19	53	2	293	28	149	16	17	2	92	15
Other	370	9	400	16	94	9	178	19	108	15	379	60
Total	4,258	100	2,536	100	1,049	100	953	100	699	100	628	100

Source	Germany		France		Brazil		UK		Other	
	[TWh]	%	[TWh]	%	[TWh]	%	[TWh]	%	[TWh]	%
Coal	306	50	30	5	10	2	138	35	1,992	29
Natural Gas	65	11	22	4	19	5	155	39	1,836	27
Petroleum	10	2	8	1	12	3	6	1	784	12
Nuclear	163	26	452	79	10	2	82	21	825	12
Other	69	11	59	11	353	88	16	4	1,376	20
Total	612	100	571	100	403	100	397	100	6,813	100

Share [%]	
USA	23
China	14
Japan	5
Russia	5
India	4
Canada	3
Germany	3
France	3
UK	2
Brazil	2
Other	36

Source: IEA statistics 2005; Author's analysis, i215

Appendix C
CO_2 Emissions by Source and Country

GLOBAL CO_2 EMISSIONS BY ENERGY SOURCES, REGION AND SECTOR

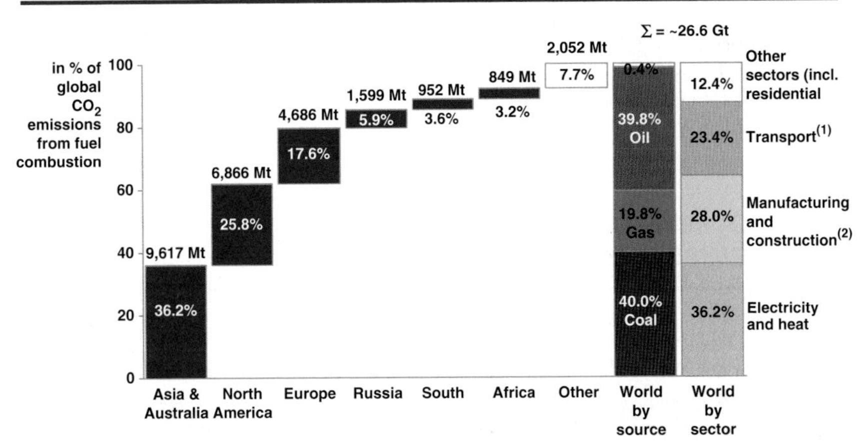

1) Transport includes Road, Marine Bunkers, International Aviation

(2) Manufacturing & Construction includes Other Energy Industries and Unallocated Autoproducers

Source: 2004 Data; IEA-CO_2 2006; Author's analysis, i302, s155

GLOBAL CO_2 EMISSIONS BY ENERGY SECTOR AND REGION

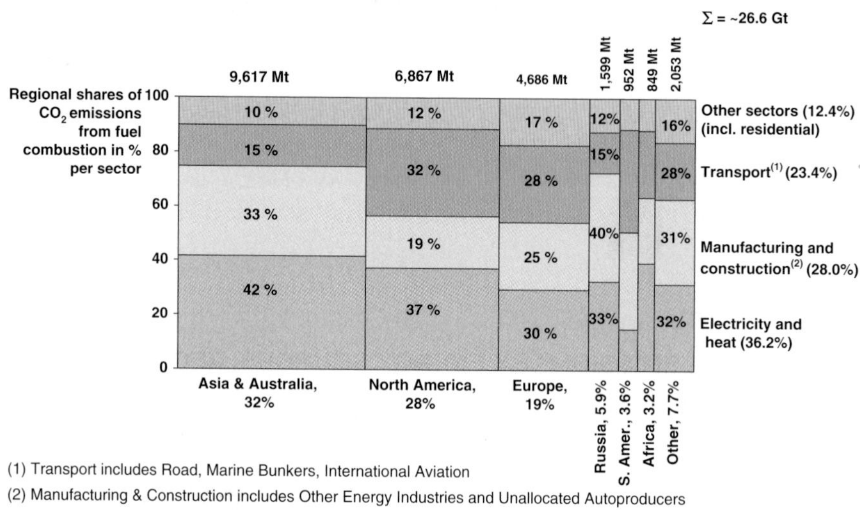

(1) Transport includes Road, Marine Bunkers, International Aviation

(2) Manufacturing & Construction includes Other Energy Industries and Unallocated Autoproducers

Source: 2004 Data; IEA-CO2 2006; Author's analysis, i308, s155

TOP 10 CO_2 EMITTORS ACCOUNT FOR 65 PERCENT OF GLOBAL EMISSIONS BY ENERGY SOURCE

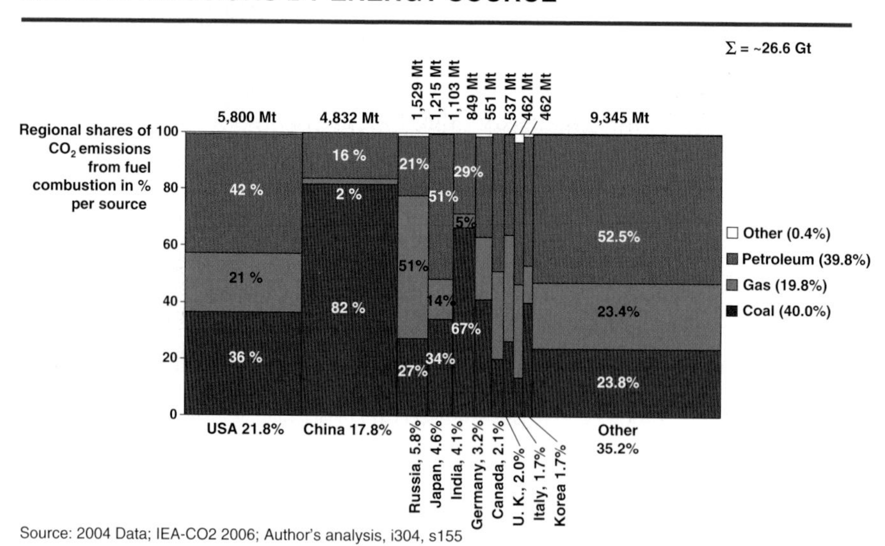

Source: 2004 Data; IEA-CO2 2006; Author's analysis, i304, s155

TOP 10 CO_2 EMITTORS 2004: COAL ONLY

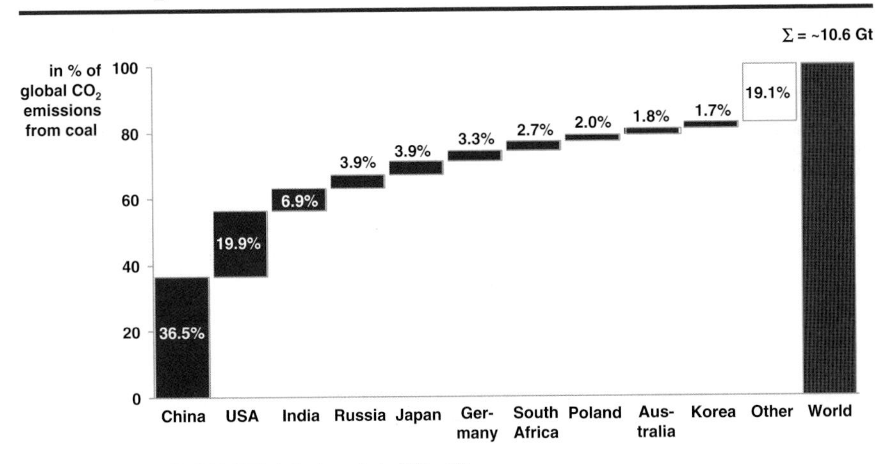

Source: 2004 Data; IEA-CO2 2006; Author's analysis, i312, s155

TOP 10 CO_2 EMITTORS 2004 FOR ELECTRICITY AND HEAT

World's CO_2 Emissions for Electricity and Heat by Energy Sources

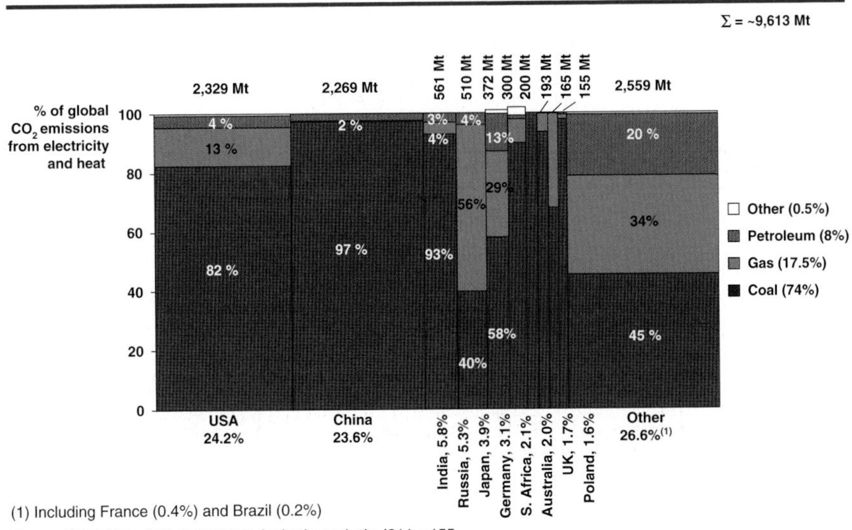

(1) Including France (0.4%) and Brazil (0.2%)

Source: 2004 Data; IEA-CO2 2006; Author's analysis, i311, s155

TOP 10 CO_2 EMITTORS 2004: TRANSPORT ONLY

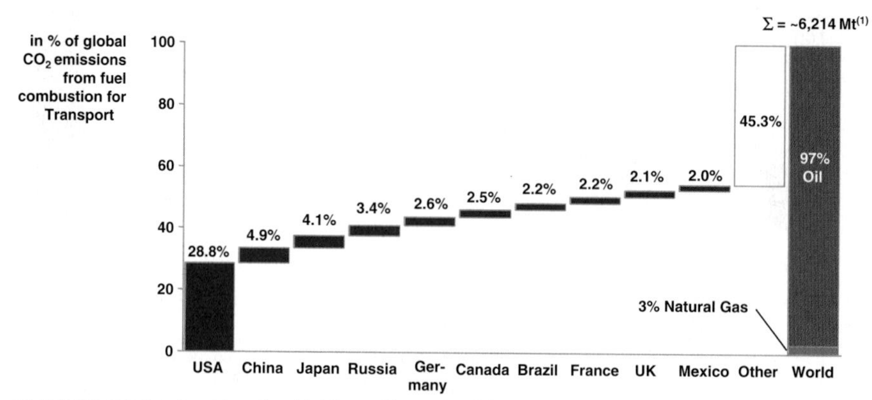

(1) Including emissions from International Aviation and International Marine Bunkers

Source: 2004 Data; IEA-CO2 2006; Author's analysis, i314, s155

The 30 power plants on the next page are the biggest CO_2-emitting power plants in the EU25 countries in absolute terms (million tons of CO_2 per year). WWF has ranked the 30 biggest emitters according to their relative emissions (Table C1).

Table C1 Dirty thirty – Europe's worst climate polluting power stations

Rank	Power plant	Country	Fuel	Start of operation	Operator	Relative emissions[a]	Absolute emissions[b]
1	Aglos Dimitrios	Greece	Lignite	1984–1986, 1997	DEH	1,350	12.4
2	Kardia	Greece	Lignite	1975, 1980–1981	DEH	1,250	8.8
3	Niederaußem	Germany	Lignite	1963–1974, 2002	RWE	1,200	27.4
4	Jänschwalde	Germany	Lignite	1976–1989	Vattenfall	1,200	23.7
5	Frimmersdorf	Germany	Lignite	1957–1970	RWE	1,187	19.3
6	Weisweiler	Germany	Lignite	1955–1975	RWE	1,180	18.8
7	Neurath	Germany	Lignite	1972–1976	RWE	1,150	17.9
8	Turow	Poland	Lignite	1965–1971, 1998–2004	BOT GiE S.A.	1,150	13.0
9	As Pontes	Spain	Lignite	1976–1979	Endesa	1,150	9.1
10	Boxberg	Germany	Lignite	1979–1980, 2000	Vattenfall	1,100	15.5
11	Belchatow	Poland	Lignite	1982–1988	BOT GiE S.A.	1,090	30.1
12	Prunerov	Czech Rep.	Lignite	1967 and 1968	CEZ	1,070	8.9
13	Sines	Portugal	Hard coal	1985–1989	EDP	1,050	8.7
14	Schwarze Pumpe	Germany	Lignite	1997 and 1998	Vattenfall	1,000	12.2
15	Longannet	United Kingdom	Hard coal	1972–1973	Scottish Power	970	10.1
16	Lippendorf	Germany	Lignite	1999	Vattenfall	950	12.4
17	Cottam	United Kingdom	Hard coal	1969–1970	EDF	940	10.0
18	Rybnik	Poland	Hard coal	1972–1978	EDF	930	8.6
19	Kozienice	Poland	Hard coal	1972–1975, 1978–1979	State owned	915	10.8
20	Scholven	Germany	Hard coal	1968–1979	E.ON	900	10.7
21	West Burton	United Kingdom	Hard coal	1967–1968	EDF	900	8.9

Table C1 (continued)

Rank	Power plant	Country	Fuel	Start of operation	Operator	Relative emissions[a]	Absolute emissions[b]
22	Fiddlers Ferry	United Kingdom	Hard coal & oil	1969–1973	Scottish & Southern	900	8.4
23	Ratcliffe	United Kingdom	Hard coal	1968–1970	E.ON	895	7.8
24	Kingsnorth	United Kingdom	Hard coal & heavy fuel oil	1970–1973	E.ON	892	8.9
25	Brindisi Sud	Italy	Coal	1991–1993	ENEL	890	14.4
26	Drax	United Kingdom	Hard coal	1974–1976, 1984–1986	AES	850	22.8
27	Ferrybridge	United Kingdom	Hard coal	1966–1968	Scottish & Southern	840	8.9
28	Großkraftwerk Mannheim	Germany	Hard coal	1966–1975,1982 and 1983	RWE, EnBW, MVV	840	7.7
29	Eggborough	United Kingdom	Hard coal	1968–1969	British Energy	840	7.6
30	Didcot A & B	United Kingdom	Hard coal & gas	1968–1975, 1996–1997	RWE	624	9.5

Source: WWF (2007).
[a]Grams of CO_2 per Kilowatt hour (g CO_2/kWh). Where two plants have the same relative emissions, the plant with the higher absolute emissions (million tons CO_2 per year) ranks as dirtier.
[b]Annual emissions for the year 2006 in million tons of CO_2.

Appendix D
Online Coal Survey Results and Questions

PARTICIPANT FUNCTIONS

Question 1: "Please specify your function regarding the coal industry."

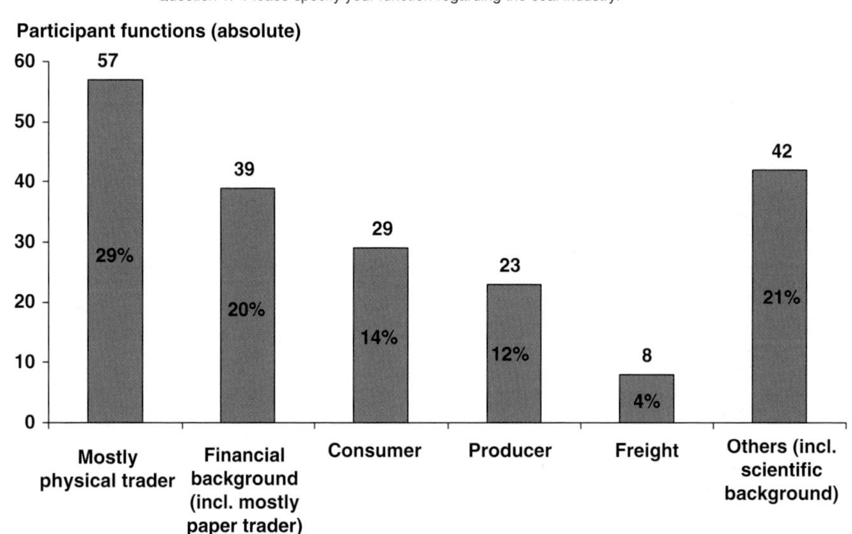

Note: n = 198

Source: Online Coal Survey: The Future of Coal–Dissertation "The Renaissance of Steam Coal" by Lars Schernikau, i401

ELECTRICITY GENERATION

Question 2: "Steam coal is mainly used to generate electric power. In the year 2005 coal provided a share of 40% of the global electricity production. In your opinion, how will this share develop until the year 2030?"

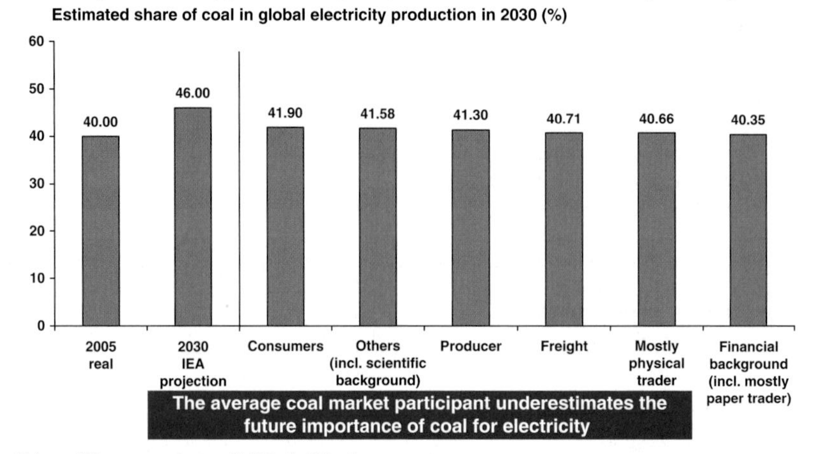

Note: n = 189, average value was 41,27% electricity share

Source: Online Coal Survey: The Future of Coal–Dissertation "The Renaissance of Steam Coal" by Lars Schernikau; i402

FUTURE PRICE DEVELOPMENTS: INTRODUCTION

Historic Coal Price Indices 2002 to 2008

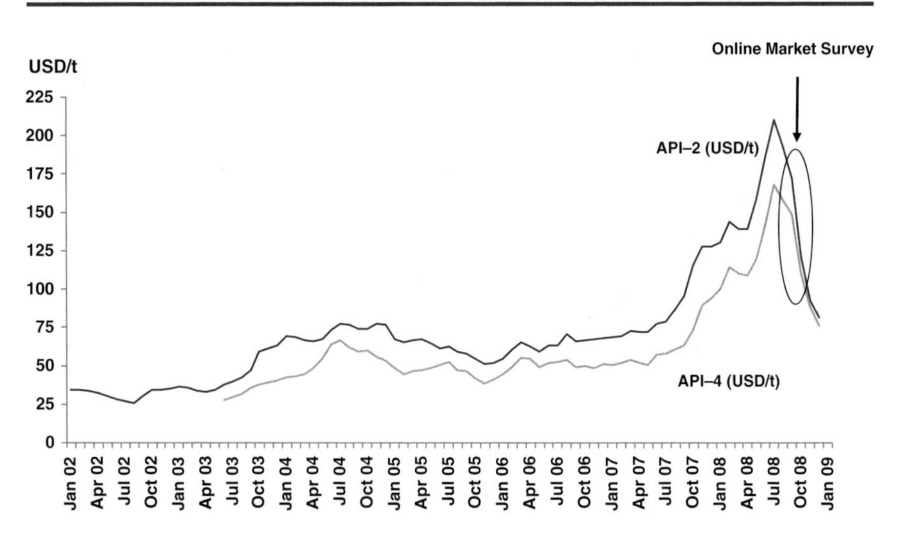

Note: API2 = coal price CIF Europe; API4 = coal price FOB South Africa

Source: McCloskey Coal Price Index, i004

FUTURE PRICE DEVELOPMENTS: SUMMARY

Participants estimated on average the following nominal coal prices for 2010

- **API2 (CIF ARA) = 157.40 USD/mt**
- **API4 (FOB RB) = 133.60 USD/mt**
- **Resulting implied freight = 23.80 USD/mt**

Participants estimated on average the following nominal coal prices for 2015

- **API2 (CIF ARA) = 174.60 USD/mt**
- **API4 (FOB RB) = 152.00 USD/mt**
- **Resulting implied freight = 22.60 USD/mt**

Participants answers correlated with the current market price movements prevailing at the time of answering the surves

- **For example, the average forecasted API2 for 2010 was around 180 USD/mt in August 2008 but only around 140 USD/mt in October 2008**

General sentiment: "However, it would be easier to win the lottery than to predict these numbers!"

Note: Questions Q5 and Q6 about future price developments; n = 157 for API2 and n = 153 for API4
Source: Online Coal Survey: The Future of Coal–Dissertation "The Renaissance of Steam Coal" by Lars Schernikau; i403

Backup

BOX PLOTS OF FUTURE PRICE DEVELOPMENTS

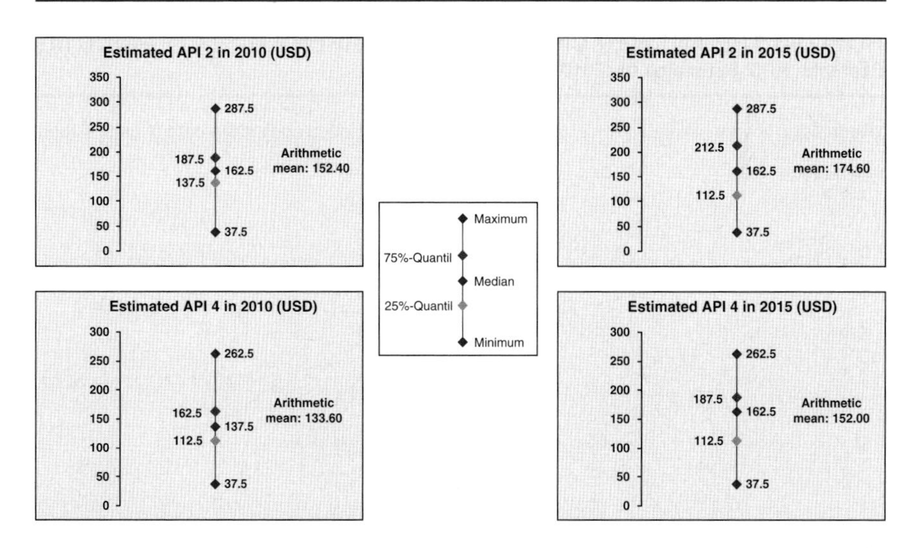

Note: Questions Q5 and Q6 about future price developments; n = 157 for API2 and n = 153 for API4
Source: Online Coal Survey: The Future of Coal–Dissertation "The Renaissance of Steam Coal" by Lars Schernikau; i404

HOW DID THE ANSWER CHANGE WITH CURRENT MARKET PRICE (I)

Example: API2 Estimates for 2010

Regression line for estimated API2 values in the year 2010, plotted against the date of participation

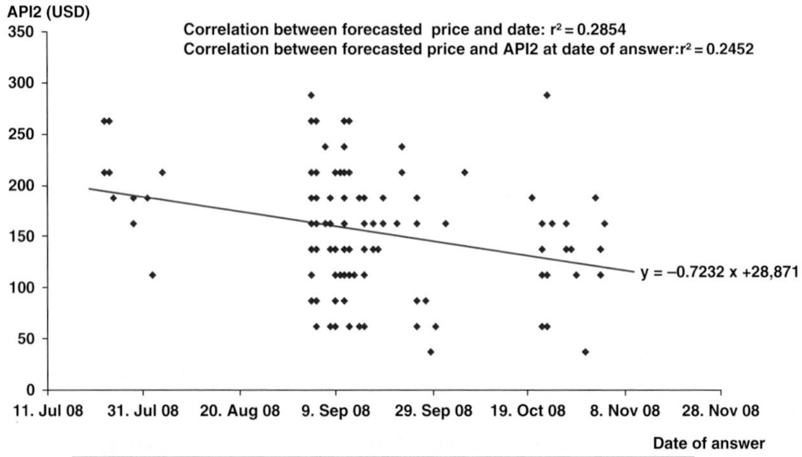

There is no strong correlation between time and price prediction, but a clear trend ...

Note: Questions Q5 and Q6 about future price developments; n = 157
Source: Online Coal Survey: The Future of Coal–Dissertation "The Renaissance of Steam Coal" by Lars Schernikau, i405

HOW DID THE ANSWER CHANGE WITH CURRENT MARKET PRICE (II)

Example: API2 Estimates for 2010

Regression line for estimated API2 values in the year 2010, plotted against the date of participation

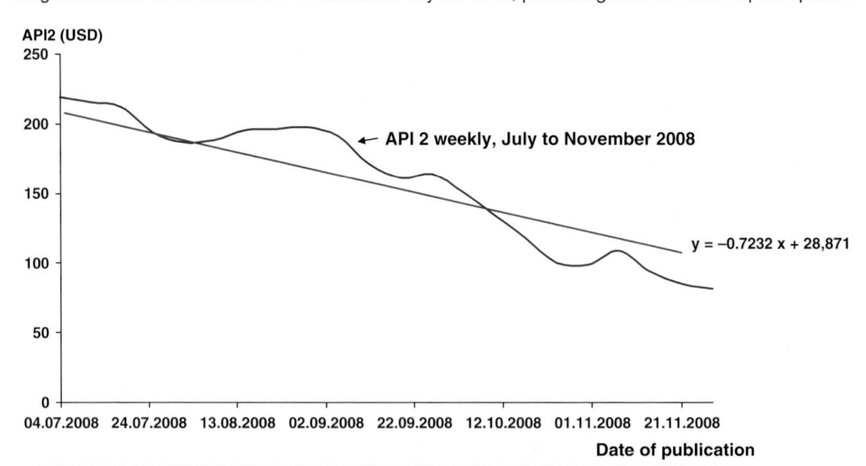

...that becomes better visible when showing the trend line on top of the real API2 price development while executing the online survey

Note: Questions Q5 and Q6 about future price developments; n = 157
Source: Online Coal Survey: The Future of Coal–Dissertation "The Renaissance of Steam Coal" by Lars Schernikau, 406

KEY DRIVERS FOR HISTORIC & FUTURE PRICE DEVELOPMENTS

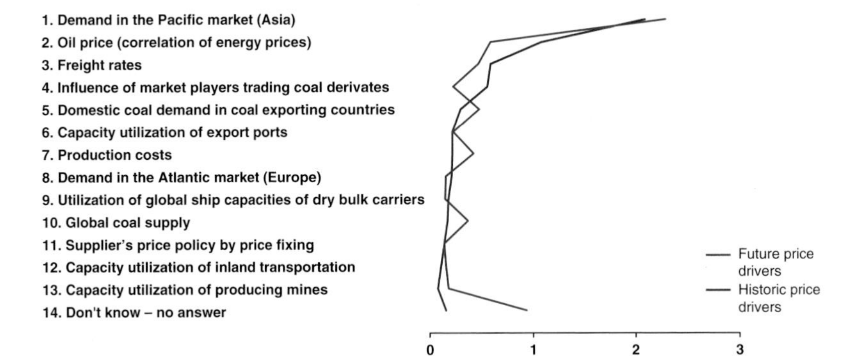

1. Points are standardized based on average (3 points for 1st place, 1 point for 3rd place)
Note: Questions Q3 and Q7 about historic and future price drivers; n=102 here only participants that forecasted higher future coal prices
Source: Online Coal Survey: The Future of Coal–Dissertation "The Renaissance of Steam Coal" by Lars Schernikau, i407

DRIVERS OF HISTORIC FREIGHT PRICE INCREASES

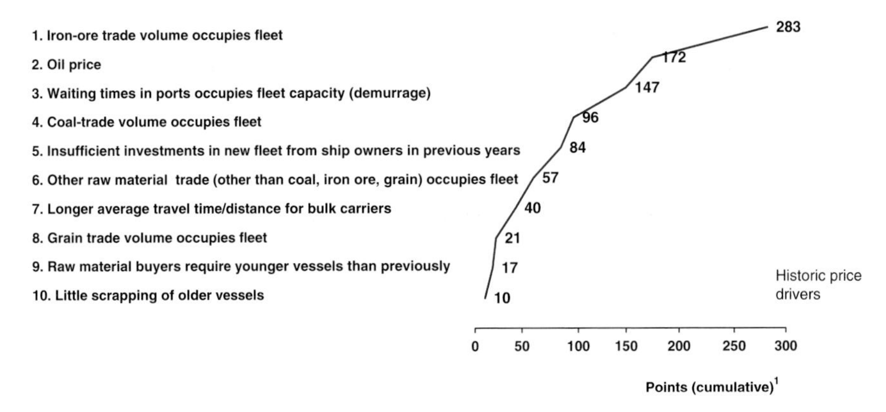

(1) Values according to ranking: 1. place: 3 points; 2. place: 2 points; 3. place: 1 point
Notes: Questions Q4 about drivers of historic freight price increases; n=189
Source: Online Coal Survey: The Future of Coal–Dissertation "The Renaissance of Steam Coal" by Lars Schernikau, i408

INFLUENCE OF COAL DERIVATIVES ON THE PHYSICAL MARKET (I)

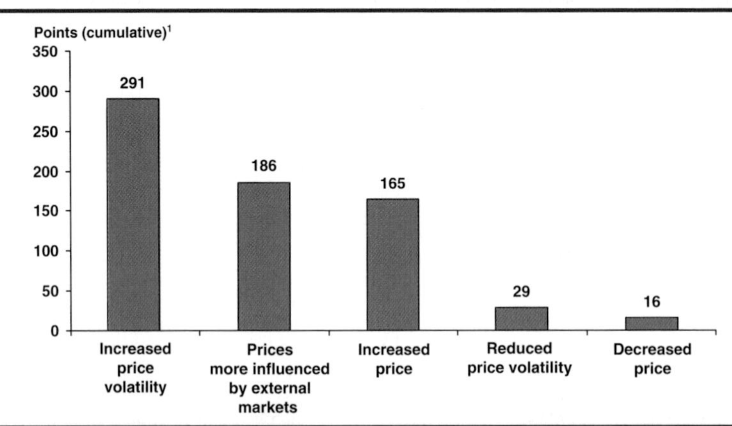

General Sentiment: "Increasing number of market participants who don't understand the main products coal/freight, increased influence of market psychology, fundamental data becomes less important, coal market develops into a stock market."

1. Values according to ranking: 1. place: 3 points; 2. place: 2 points; 3. place: 1 point
Note: Questions Q9 about influence of coal derivatives; n = 140
Source: Online Coal Survey: The Future of Coal–Dissertation "The Renaissance of Steam Coal" by Lars Schernikau, i409

INFLUENCE OF COAL DERIVATIVES ON THE PHYSICAL MARKET (II)

Answer distribution by function

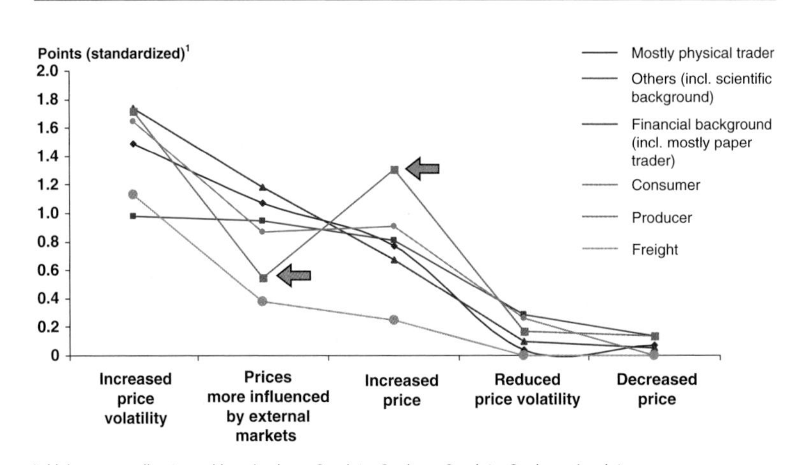

1. Values according to ranking: 1. place: 3 points; 2. place: 2 points; 3. place: 1 point
Note: Questions Q9 about influence of coal derivatives; n = 140
Source: Online Coal Survey: The Future of Coal–Dissertation "The Renaissance of Steam Coal" by Lars Schernikau, i410

MARKET CONCENTRATION, RISK OF "COAL-PEC"

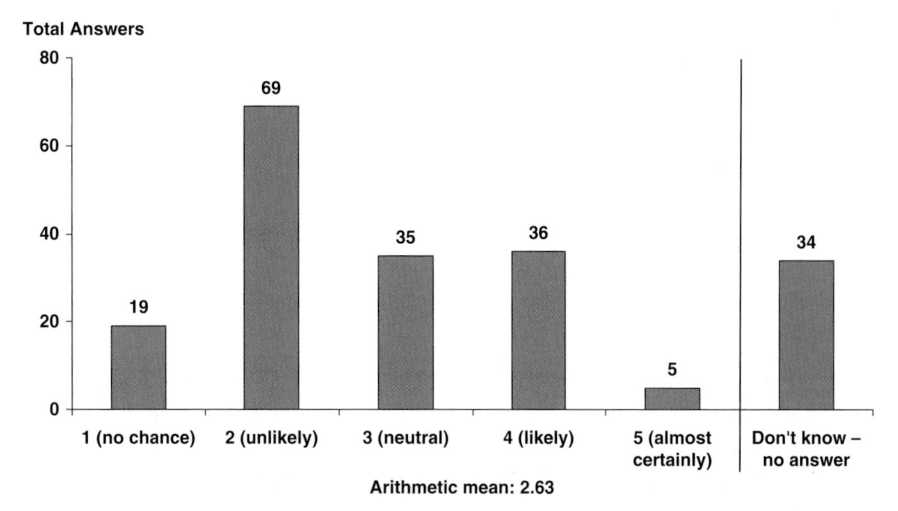

Note: Questions Q10: In your opinion, is there a chance or a risk that a "Coal-OPEC" will emerge within the next 5–10 years? N = 134, Standard Deviation 1,05, Mean 2,63
Source: Online Coal Survey: The Future of Coal–Dissertation "The Renaissance of Steam Coal" by Lars Schernikau, i411

MOST INFLUENTIAL PRODUCERS & TRADERS IN STEAM COAL MKT.

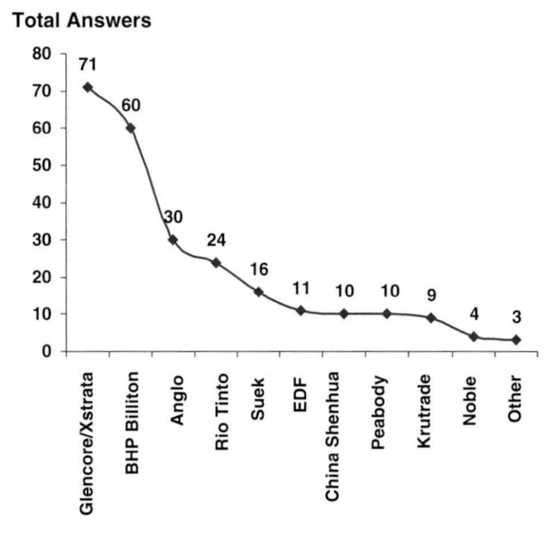

Rank	Producer/Trader	Total answers
1	Glencore/Xstrata	71
2	BHP Billiton	60
3	Anglo (Coal)/Anglo American	30
4	Rio Tinto	24
5	Suek	16
6	EDF	11
7	China Shenhua Energy	10
8	Peabody	10
9	Krutrade	9
10	Noble	4
11–16	EON, Gazprom, Constellation, Cargill, Weglokoks, CMC/Carbocoal	3 each

Note: Questions Q11: List and -if possible-rate the producers and traders, you regard as the most influential (1 = most important); n = 198
Source: Online Coal Survey: The Future of Coal–Dissertation "The Renaissance of Steam Coal" by LarsSchernikau, i412

CHINA'S EXPECTED IMPORT-EXPORT-BALANCE IN 2020

Development of China's import-export balance in 2020

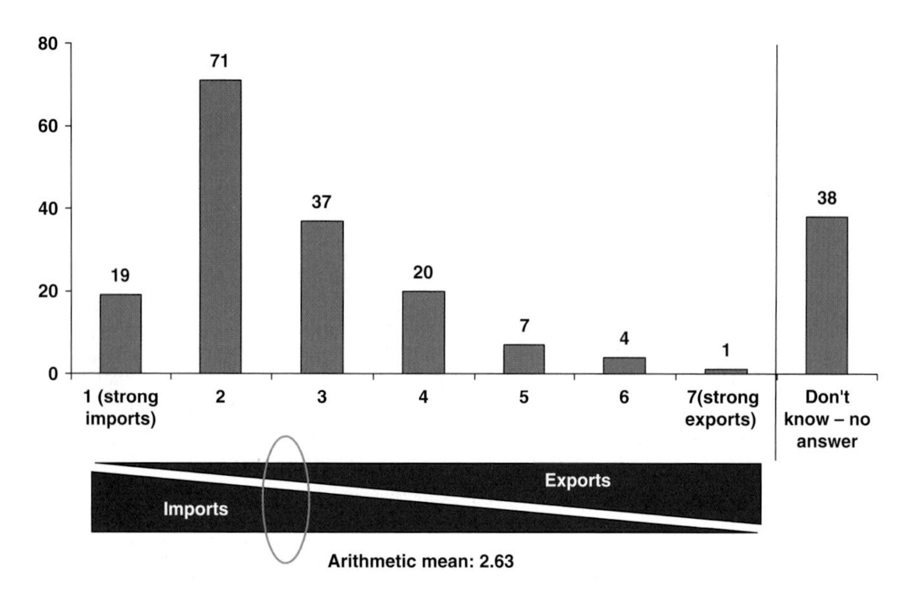

Note: Questions Q12: In your opinion how will China's Import-Export-Balance develop until the year 2020? Please rate from 1(much import-no export) to 7 (much export -no import); n = 198
Source: Online Coal Survey: The Future of Coal–Dissertation "The Renaissance of Steam Coal" by Lars Schernikau, i413

ENERGY SCENARIOS: COAL SCENARIO MOST LIKELY

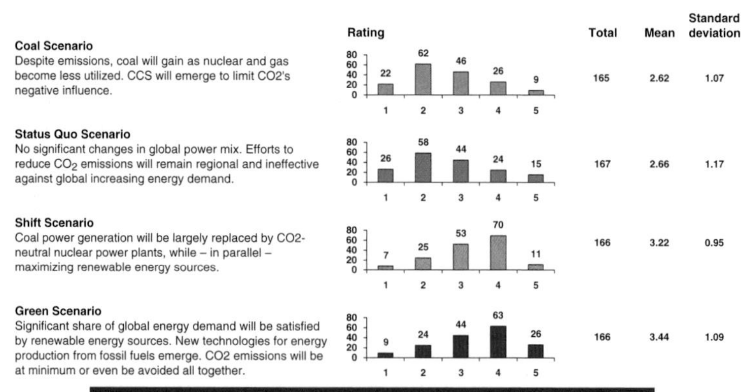

Average coal market participant expects that Green Scenario is less realistic within 20 years but comments indicate hope for beyond 2050

Notes: Questions Q13: Please rate the following energy scenarios for 2030, Rating from "very likely"(1) to "no chance at all"(5):; n = 167
Source: Online Coal Survey: The Future of Coal–Dissertation "The Renaissance of Steam Coal" by Lars Schernikau, i415

CCS TECHNOLOGY EXPECTED ECONOMICAL IN NEAR FUTURE

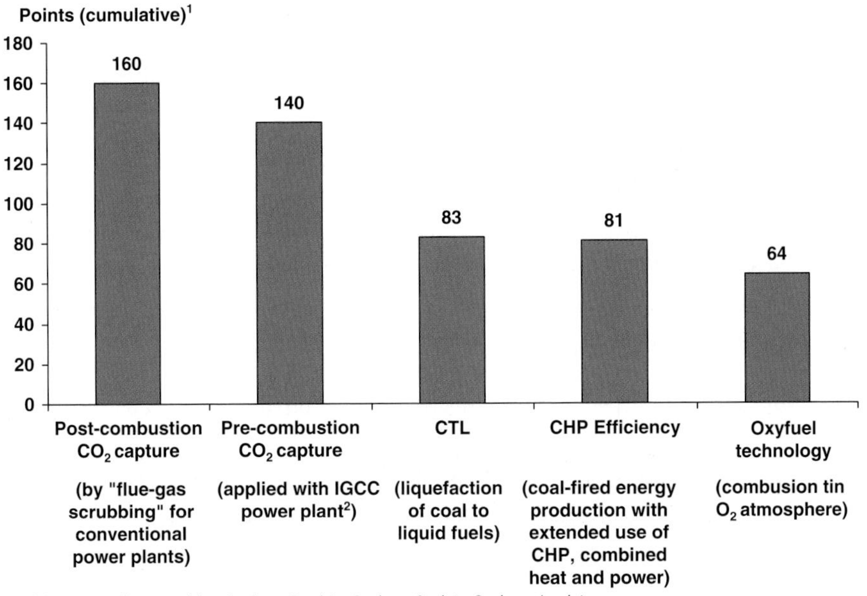

1. Values according to ranking: 1. place: 3 points; 2. place: 2points; 3. place: 1 point
2. Integrated Gasification Combined Cycle

Notes: Questions Q14: Which of the following technological processes has in your opinion best chances to be economically used in the nearest future? ; n = 109

Source: Online Coal Survey: The Future of Coal–Dissertation "The Renaissance of Steam Coal" by Lars Schernikau, i416

Online Coal Survey Questions, Word Format (14 Questions)

Dear Participant,

Secure energy supply is one of the most important issues for the world's future. The use of coal for producing energy generates versatile positions in politics, society and industry. With its high greenhouse gas emissions compared to other fossil fuels, the combustion of coal is likely to be one of the main reasons for global warming. On the other hand coal is the most available fossil energy resource with a statistical range of more than 150 years. As a part of a dissertation at the Institute of Energy Systems at Technische Universität Berlin we have analyzed the future of coal regarding its potential in prospective use and the structure of the coal market. This survey only covers the commodity steam coal. Other sorts of coal are excluded.

It would be very important for us that you, as an expert of your area, offer us some of your valuable time to answer our questions on the future of coal.

Question 1: Please specify your function regarding the coal industry. Max 2 Answers

Producer; Trader (mostly physical); Trader (mostly paper); Consumer; Engineer (with consumer); Engineer (with producer); Function with a financial background (banks, financier, investor, etc.); Function in the energy market, other than coal; Researcher/Scientist; Politician; Other (i.e. outside of energy market and none of the above). Please name under 'Further remarks on this question'.

Question 2: Electric Power Generation – Steam coal is mainly used to generate electric power. In the year 2006 coal provided a share of 40% of the global electricity production. In your opinion, how will this share develop until the year 2030?

- Decrease substantially (i.e. to 35% or lower)
- Decrease (i.e. to below 40%, but above 35%)
- Stay constant
- Increase (i.e. to above 40%, but below 45%)
- Increase substantially (i.e. to above 45%)
- Don't know/no answer

Question 3: Historic Coal Price Drivers – The price index for traded steam coal in Europe (API 2) averaged ~US $88.5/ton in the year 2007. Compared to the year 2006 this meant an increase of more than US $24.5/ton, or 38%. In the year 2008 prices increased further.

Mention maximum three reasons indexed by importance (1 = most important) which caused that increase as well as current price hikes. Please choose from the list below.

- Decrease in global coal supply
- Increasing demand in the Atlantic market mainly by Europe

- Increasing demand in the Pacific market mainly by Asia
- Increasing demand in the coal exporting countries
- Increase in freight rates
- Utilization of global ship capacities of dry bulk carriers
- Capacity utilization of exports ports
- Capacity utilization of inland transportation
- Capacity utilization of producing mines
- Increase of production costs
- Influence of market players trading coal derivates
- Influence of the increased oil price (correlation of energy prices)
- Supplier's price policy by price fixing
- Other. Please name under 'Further remarks on this question'

***Question 4*:** Historic Freight Price Drivers – One of the reasons for the increased price level of imported steam coal around the world is the freight price increase for seaborne dry bulk carriers. Compared to the year 2006 the price for Panamax-sized freight from South Africa to Europe has almost doubled from US $15 to 30/ton in 2007. In the year 2008 prices increased further.

Mention maximum three reasons indexed by importance (1 = most important) which caused that increase as well as current price hikes. Please choose from the list below.

- Increase in iron ore trade volume occupies fleet
- Increase in grain trade volume occupies fleet
- Increase in coal trade volume occupies fleet
- Increase in other raw material trade (other than coal, iron ore, grain) occupies fleet
- Raw material buyers require younger vessels than in previous years
- Too much scrapping of older vessels
- Longer waiting times in ports occupies fleet capacity (demurrage)
- Longer average travel time/distance for bulk carriers
- Insufficient investments in new fleet from ship owners in previous years
- Oil price increase
- Other. Please name under 'Further remarks on this question'

***Question 5*:** Future FOB Prices API4 – How will the index API4 (Free on board, Richards Bay, South Africa) develop in your opinion? Please estimate the range of the nominal prices at (a) the end of the year 2010 and (b) at the end of the year 2015. Please estimate as well as you can, even though the information available is limited.

Choose from one of the following price ranges each for 2010 and 2015:

less than $50/ton; 50–75 $/t; 75–100 $/t; 100–125 $/t; 125–150 $/t; 150–175 $/t; 175–200 $/t; 225–250 $/t; more than 250 $/t

***Question 6*:** Future CIF Prices API2 – How will the index API2 (Cost, Insurance, Freight; Amsterdam, Rotterdam, Antwerpen) develop in your opinion? Please estimate the range of the nominal prices at (a) the end of the year 2010 and (b) at the end of the year 2015. Please estimate as well as you can, even though the information available is limited.

Choose from one of the following price ranges each for 2010 and 2015:

Less than \$50/ton; \$50–75/ton; \$75–100/ton; \$100–125/ton; \$125–150/ton; \$150–175/ton; \$175–200/ton; \$225–250/ton; \$250–275/ton; more than \$275/ton

***Question 7*:** Future Price Drivers – For the year 2010 you forecast constant, decreasing, or increasing price levels for seaborne traded steam coal. What reasons could this development have?

Mention maximum three reasons indexed by importance (1=most important) which will cause future price increases. Please choose from the list below.

- Decrease in global coal supply
- Increasing demand in the Atlantic market mainly by Europe
- Increasing demand in the Pacific market mainly by Asia
- Increasing demand in the coal-exporting countries
- Increase in freight rates
- Utilization of global ship capacities of dry bulk carriers
- Capacity utilization of exports ports
- Capacity utilization of inland transportation
- Capacity utilization of producing mines
- Increase of production costs
- Influence of market players trading coal derivates
- Influence of the increasing oil price (correlation of energy prices)
- Supplier's price policy by price fixing
- Other. Please name under 'Further remarks on this question'.

Note: Questions 7 and 8 where similar, 7 was to be answered for people who assumed increasing future prices and the other was for people that assumed decreasing future prices. Here we only show Question 7 for participants who chose increasing price levels.

***Question 8*:** Coal Derivates – The paper trading with coal derivates/coal-related securities for the European import market increased within 1 year (2006–2007) by more than 55% to 1.4 billion tons (only API2 derivates). In the same period the physically traded steam coal imported to Europe amounted to ~250 million tons. This corresponds to a 1:6 ratio. Which in your opinion is the influence of the paper market on the physically traded steam coal market?

Please choose from the list below a maximum of three answers indexed by importance (1 = most important).

- Reduction of price volatility
- Intensification of price volatility

- Steam coal prices get under increased influence of external markets
- Increase of the physical steam coal's market price
- Decrease of the physical steam coal's market price
- No influence
- Other. Please name under 'Further remarks on this question'.

***Question 9*:** Market Concentration – The world market for steam coal increasingly focuses on a small number of very large producers and traders. In the year 2006, nine major producers/exporters represented 60% of the global seaborne steam coal market. In your opinion, is there a chance or a risk that a 'Coal-PEC' will emerge within the next 5–10 years?

1 (no chance); 2 (unlikely); 3 (neutral); 4 (likely); 5 (almost certainly); Don't know/no answer

***Question 10*:** Most Influential Companies

List and – if possible – rate the producers and traders, you regard as the most influential (1 = most important).

***Question 11*:** China – At the Coaltrans 2007 in Rome 43% of the participants agreed that the thermal coal export of China will disappear by 2020. In your opinion how will China's import–export balance develop until the year 2020?

Please rate from 1 (much import – no export) to 7 (much export – no import)

***Question 12*:** Energy Scenarios – Regarding the future of global energy supply (outlook for the year 2030), different scenarios resulting in different output levels of greenhouse gas emissions are discussed:

- Green Scenario: A significant share of global energy demand will be satisfied by renewable energy sources. New technologies for energy production from fossil fuels will emerge. As a result, CO_2 emissions will be at minimum or even be avoided all together.
- Shift Scenario: Coal power generation will be largely replaced by CO_2-neutral nuclear power plants, while – in parallel – maximizing renewable energy sources.
- Status Quo Scenario: There won't be significant changes in the global power mix. Efforts to reduce CO_2 emissions will remain a regional rather than a global effort and will be ineffective against the background of global increasing energy demand.
- Coal Scenario: Notwithstanding the emissions, coal will gain in relative attractiveness as nuclear and gas become less utilized. CCS (Carbon Capture and Storage) will emerge to limit CO_2's negative influence.

In your opinion how likely are the above-described scenarios to become reality? Please rate from 'very likely' to 'no chance at all.'

Question 13: Coal Technologies – A number of technologies are currently being discussed concerning the future of coal. In addition to more efficient combustion, it also becomes possible to capture and store the emerging CO_2. Another idea would be to utilize steam coal increasingly in application areas that were largely ignored until now.

Which of the following processes has in your opinion chances to be economically used in the nearest future?

Please create a list in order of highest likelihood (1=most likely). Please rank the six elements below:

- High-efficient coal-fired energy production with extended use of CHP (Combined Heat and Power)
- Post-combustion CO_2 capture by 'flue gas scrubbing' for conventional power plants
- Oxyfuel technique (combustion in O_2 atmosphere)
- Pre-combustion CO_2 capture applied with IGCC power plant (Integrated Gasification Combined Cycle)
- CTL (liquefaction of coal to liquid fuels)
- Other. Please name under 'Further remarks to this question.'
- Don't know

Thank you for your participation. We are very grateful for your help. As soon as we have evaluated the results of this Online Survey we will send you a summary.

Appendix E
WorldCoal Market Model

WorldCoal is a nonlinear model to quantitatively analyze the global steam coal trade. The model enables real-time analysis of the current market situation, which was conducted for the year 2006 as described here. With this model, it is possible to analyze certain likely future scenarios, such as production capacity or logistical constraint scenarios. WorldCoal was developed in 2008 as part of Stefan Endres' research project, coached and supervised by the author and Professor Georg Erdmann, and the Department of Energy Systems at the Technical University in Berlin. WorldCoal models CIF cost and price levels in 2006, thus extending beyond the previously discussed FOB cost levels.

In this section the following variables are used.

a	variable for the value of a curve where $q = 0$
b	variable for the incline of a function
Π	profit
$C = \mathrm{TC}$	total costs
C_{f}	fixed costs
c_{v}	specific variable costs
CIT	GAMS: costs inland transportation (including transshipment)
COST	GAMS: objective function
CPR	GAMS: production costs
CTS	GAMS: freight costs
d	distance
dem	demand, i.e., for demand quantity
i	index supplier
j	index consumer
MC	marginal cost
MS	marginal supplier
n	time interval
p	price
q	quantity of one market player
Q	total quantity
q_{s}	supplied quantity
r	annual rate of increase

supmax	maximum supply
supmin	minimum supply
t	end of forecast time
t_0	beginning of forecast time

Methodology, Definitions, and Mathematical Structure of WorldCoal

The WorldCoal model uses GAMS programming (General Algebraic Modeling System) to optimize the distribution of the 2006 coal supply in such a way that the total cost (TC) of the market is minimized. A maximum supply for each supplier was defined, based on production or logistical constraints. These constraints were calibrated with real-life constraints in the various markets for the year 2006, such as in South Africa (as discussed in previous chapters).

Modeling of the global steam coal market corresponds to solving a transportation problem. Rosenthal 2008 describes in the *trnsport.gms* model a basic market with two producers and three consumers. WorldCoal extended that model. GAMS optimizes transportation for the total cost minimum, given the constraints of the producers (supmax and supmin) and demand by consumers (dem). Today, there are many examples of complex models for determining energy market equilibriums. For instance, Holz/Hirschhausen/Kemfert (2008) modeled the European gas market in their model GASMOD. When modeling the steam coal market, I utilized much of the systematic of GASMOD but adjusted appropriately for the coal market (Fig. E1).

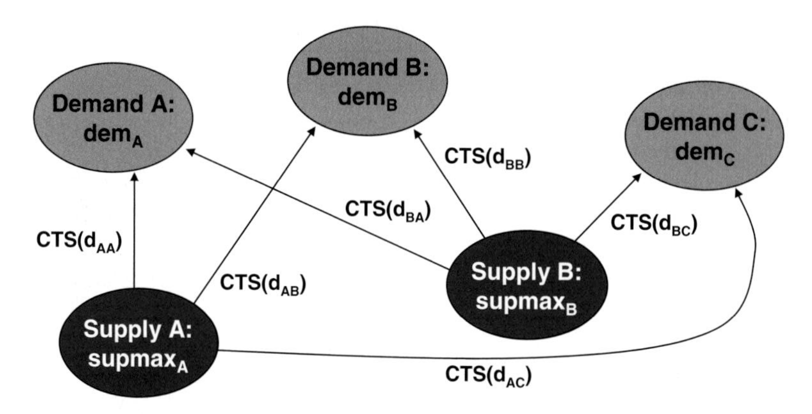

Fig. E1 The transportation problem solved with *GAMS* (Source: Endres 2008)

As already discussed in Section 5.4, supply is divided into various supply regions. The same is true for demand. In the WorldCoal model the regions were slightly adjusted compared to the FOB supply curve regions discussed in the previous sections. Russia was considered as two supply regions, one for the Atlantic (west) and the other for the Pacific (east). In total, the model differentiates between eight supply regions and nine demand regions as detailed in Table E1.

For the purpose of the WorldCoal model, the seaborne steam coal market is considered to be perfectly competitive. Thus, no one single supply region or demand

Table E1 WorldCoal model's geographic definitions

Exporters			Importers		
Country	**Abbr.**	**Port**	**Country**	**Abbr.**	**Port**
Australia	AU	Newcastle	Europe	EU	ARA
South Africa	SA	Richards Bay	Japan	JA	Yokohama
Indonesia	ID	Banjarmasin	Korea	KR	Ulsan
Russia-Poland West	RU_WEST	Riga	Taiwan	TW	Kaohsiung
Russia-Poland East	RU_EAST	Vostochny	China	CH_IM	Hong Kong
Colombia-Venezuela	CO	Puerto Bolivar	India	IN	Mundra
China	CH-EX	Rizhao	Latin America	LA	Acapulco
USA-Canada	US_EX	Jacksonville	USA-Canada	US_IM	Mobile
			Asia Other	AS_OT	Manila

Sources: Endres (2008) and Pinchin-Lloyd's (2005)

region can influence price or quantity; rather, they have to take these as given (Varian 1999). The individual regions' supply and demand functions add up to a worldwide supply and demand function. I also assumed this principle when building the FOB supply curve in Section 5.4. This assumption is reasonable for the global supply curve, especially when taking a single supply region as a competitor. No one single supply region has any market power, just as no one single demand region has any market power. In perfect competition, the market price equals the marginal cost of the marginal supply (see also the price discussion in Section 5.3.4).

I will refer to 'supplier' below for simplicity, which I use synonymously with 'supply region'.

Objective Function: The nonlinear WorldCoal model consists of a nonlinear objective function and a number of linear restrictions. The model minimizes total costs TC. Here i is the index for all suppliers and j is the index for all consumers, the objective function is defined in formula (E1). The quantity q_{ij} signifies the quantity of coal that is shipped from producer i to consumer j.

$$\min \sum_{i} \sum_{j} (\mathrm{TC}_{ij}) q_{ij} \tag{E1}$$

Total cost was already defined as the sum of freight costs CTS from producer i to consumer j, inland and transshipment costs CIT plus production costs CPR.

$$\mathrm{TC}_{ij} = \mathrm{CTS}_{ij} + \mathrm{CIT}_i + \mathrm{CPR}_i \tag{E2}$$

$$\mathrm{CPR}_i = a_i + b_i q_{\mathrm{s},i} \tag{E3}$$

$$q_{s,i} = \sum_i q_{ij} \tag{E4}$$

Formula (E4) simply states that the total produced/shipped quantity q_s of supplier i equals the sum of all individual shipped quantities q_{ij} from supplier i to all consumers j.

Here $b > 0$, the objective function (E1) can now be rewritten as a quadratic objective function as in formula (E5) below.

$$\begin{aligned} &\min \sum_i \sum_j \left(\mathrm{CTS}_{ij} + \mathrm{CIT}_i + \left(a_i + b_i \sum_j q_{ij} \right) \right) q_{ij} \\ &\Leftrightarrow \min \sum_i \sum_j \left(\mathrm{CTS}_{ij} + \mathrm{CIT}_i + a_i + b_i \sum_j q_{ij}^2 \right) \end{aligned} \tag{E5}$$

Restrictions: Supmax_i is defined as the maximum supplied quantity of supplier i considering shipments to all consumer regions j.

$$\sup \max_i = \sum_j q_{ij} \tag{E6}$$

The total market demand is defined as the sum of all demanded quantities of all consumer regions $\sum_j \mathrm{dem}_j$. In market equilibrium, dem_j equals the sum of all individual quantities of the various suppliers i to that one demand region j, while considering the objective function to minimize total cost in the market. Restriction function (E7) therefore implicitly assumes that there is always enough coal to satisfy the entire demand and the consumers basically accept any price that results from the market equilibrium.

$$\mathrm{dem}_j = \sum_i q_{ij} \tag{E7}$$

Pricing: As discussed previously, the equilibrium price in perfect competition equals the marginal cost of the most expensive or marginal supplier to region j $\mathrm{MC}_{\mathrm{MS}(j)}$. Formula (E8) below is used to calculate the price for each demand region j.

$$p_j = c_{v,\mathrm{MS}(j)} = \mathrm{MC}_{\mathrm{MS}(j)} \tag{E8}$$

The market price can therefore be determined by calculating the derivative of the objective function.

$$p_j = \mathrm{MC}_{\mathrm{MS}(j)} = \frac{\partial \mathrm{COST}_{\mathrm{MS}(j)}}{\partial q_{\mathrm{MS}(j)}} \tag{E9}$$

Input Data

FOB Costs: The proprietary FOB marginal cost data from Section 5.4 have been used to fill the WorldCoal model with cost data. For the global supply function, I could approximate an exponential supply cost function as depicted in Fig. 5.13 for each supply region; however, I do not possess that detail for every supply region. Thus, the individual marginal cost data of the sub-supply regions have been aggregated into bended supply functions. The model requires an approximation to build the relationship between FOB costs and the offered quantity q_s. Therefore, we separated the supply function for each supply region into two parts: part (1), which has a gentler slope, and part (2), which has a steeper slope as we get closer to the supmax_i quantity. This is consistent with the real world, where the marginal cost increases at a faster rate, the closer we get to the maximum possible output. Thus supply is less elastic the closer we are to the maximum output quantity.

The approximated two-step supply function for each supply region is therefore nonlinear and consists of two different linear functions.

$$q_{s,i} = \left\{ \begin{array}{l} FOB_{1,i} = a_{1,i} + b_{1,i} q_{s1,i}\,; \; \left[0 \le q_{s1,i} \le \text{supmax}_{1,i}\right] \\ FOB_{2,i} = a_{2,i} + b_{2,i} q_{s2,i}\,; \; \left[\text{supmax}_{1,i} \le q_{s2,i} \le \text{supmax}_{2,i}\right] \end{array} \right\} \qquad \text{(E10)}$$

The Endres 2008 paper details all supply regions' functions utilized in the WorldCoal model. Figure E2 illustrates an example for the supply region 'Russia/Poland.' Also inserted into the figure are the quantities that WorldCoal calculated for 2006 and the 2015 forecast quantity.

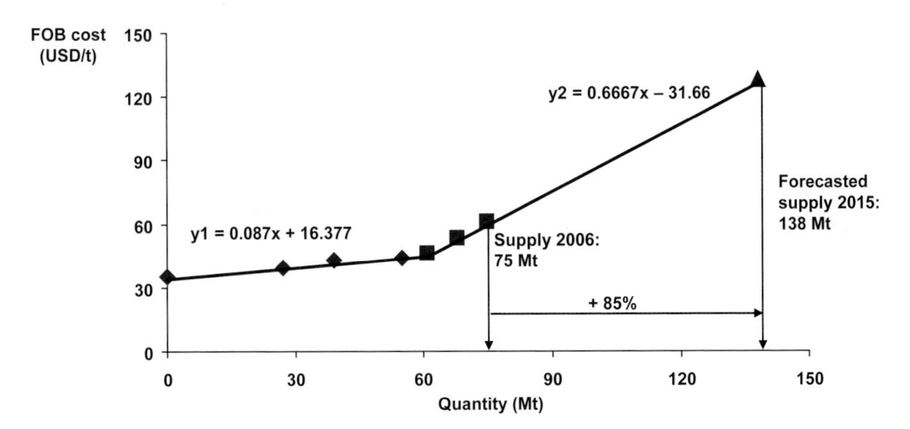

Fig. E2 WorldCoal's approximated supply function Russia/Poland (Source: Author's FOB marginal cost analysis; Endres 2008)

Freights: The freight problem in WorldCoal was approximated by allocating one loading port per supply region and one discharging port per demand region. The ports and their respective region were summarized in Table E1. The actual freight rates are published regularly in various industry publications. Freight data from Pinchin-Lloyd's (2005) were taken and a linear freight rate function approximated, where freight rates depend on distance. The distance between each pair of supplier and consumer is defined as d.

$$\mathrm{CTS}(d) = 6.5 + 0.002d \tag{E11}$$

Results for the Reference Year 2006

For each supply and demand region, WorldCoal's results consist of quantity data, CIF, and FOB prices, as well as sensitivity data that can be used to evaluate What–If analyses.

$$\mathrm{CIF}_{ij} = \mathrm{CPR}_i + \mathrm{CIT}_i + \mathrm{CTS}_{ij} = \mathrm{FOB}_i + \mathrm{CTS}_{ij} \tag{E12}$$

WorldCoal's CIF costs, as summarized in Table E2, are the theoretical import price for each region. The CIF costs are the sum of the marginal suppliers' marginal FOB costs and freight. The values that are actually relevant for WorldCoal's supply to a region are marked.

Table E2 WorldCoal 2006 CIF costs

WorlCoal CIF- Costs (US$/ton)

	EU	JA	KR	TW	CH_IM	IN	LA	US_IM	AS_OT
AU	75	60	60	60	60	64	65	70	60
SA	65	66	66	63	63	58	68	67	60
ID	59	46	46	44	44	47	58	64	45
RU	69	69	68	70	71	79	80	78	73
CO	60	68	69	71	71	69	55	54	74
CH_EX	71	52	51	51	68	60	64	70	54
US_EX	58	69	70	71	72	68	56	73	72

Sources: WorldCoal; Endres (2008).

WorldCoal Quantities

In WorldCoal's 2006 optimum, quantities are distributed as summarized in Table E3. The optimum shows that Japan (JA) is supplied only by Australia (AU). Taiwan (TW), China (CH_IM), and other Asia (AS_OT) are supplied by the worldwide lowest marginal cost supplier, Indonesia (ID). Because of their geographical proximity, Colombia (CO) supplies to Latin America (LA) and the United States (US_IM).

Table E3 WorldCoal 2006 quantity allocation

Coal quantity allocation (million tons)

	EU	JA	KR	TW	CH_IM	IN	LA	US_IM	AS_OT	Supply (million t)
AU		114								114
SA	59					9				68
ID			2	54	50	19			46	171
RU	75									75
CO	18						11	40		69
CH_EX			59							59
US_EX	39									39
Demand (million t)	191	114	61	54	50	28	11	40	46	**SUM 595**

Sources: WorldCoal; Endres (2008).

Australia exhausted its entire quantity, but Indonesia and Colombia still have some unsold coal. Table E2 shows that Indonesia should then supply South Korea (KR) and India (IN) to optimize costs. However, because of the defined restrictions, there is not enough quantity available to satisfy demand for both. For India, the next cost optimal supplier is South Africa (SA) and for South Korea this would be China (CH_EX). China, as the alternative supplier with the lower costs, now first delivers its entire supply to South Korea. The remaining demand of South Korea is satisfied by Indonesia. India only then receives the remaining quantity available from Indonesia and satisfies its remaining demand with coal from South Africa.

The same systematic can then be used for the consumer region Europe (EU). The market is now in equilibrium. Supply equals demand. GAMS found the equilibrium in such a way that the total costs (objective function COST) of the entire market are minimized. Please note that the real-world coal market is significantly more complex and the real quantities may differ from the calculations. The model, however, helps to understand the real world by simplifying it to a manageable complexity.

The extent to which the real quantity distribution differed from WorldCoal's calculations is summarized in Fig. E3. The model concentrates the volumes supplied to each consumer region on fewer supply regions. The real quantities delivered show a larger supply fragmentation. That is expected since the simplified model does not account for the various intricacies of the market discussed in Section 5.2. However, the model very closely predicted deliveries to the complex demand region of Europe and forecast the correct supply tendencies for all other demand regions other than Latin America, which in fact is an insignificant coal user in the real world anyways.

WorldCoal Prices

The market equilibrium is fully determined when not only quantities but also prices are in equilibrium. In a perfectly competitive market, price equals the marginal cost

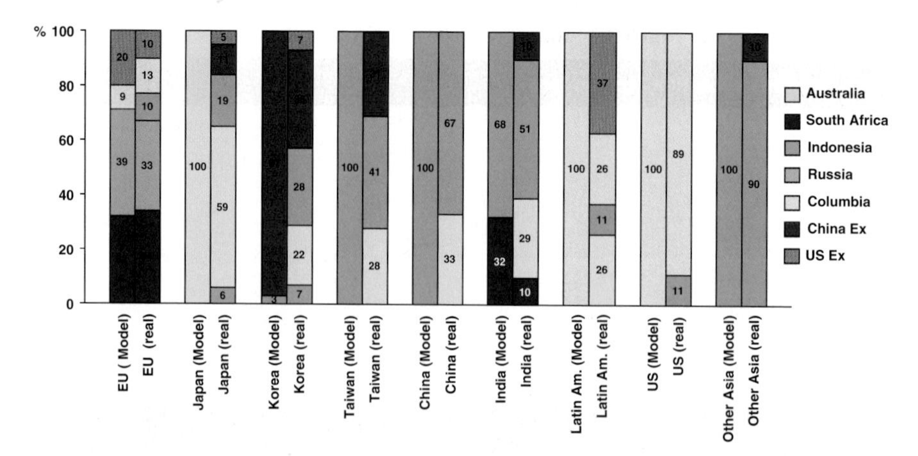

Fig. E3 WorldCoal's and real relative quantity allocation 2006 (Source: WorldCoal; Endres 2008; Author's analysis)

of the marginal supplier. Thus, it seems that prices should equal the CIF costs of the marginal supplier for a region, as detailed in Table E2. However, WorldCoal models more than one supply and one demand region. Thus, the pure CIF costs of the marginal supplier for a region will not necessarily equal the market price in that region because the CIF costs do not account for the opportunity cost of a supplier that may be lost by not delivering to another consumer. In fact, the marginal (or best) opportunity cost of the marginal supplier for a region has to be considered because he will have to earn the fair price he could earn by supplying another consumer. This interpretation is coherent with the definition of marginal cost in competitive market price theory. The ShadowPrice is therefore introduced as follows:

$$\text{Marginal Cost} + \text{Shadow Price} = \text{Market Price} \tag{E13}$$

In linear and nonlinear programming, shadow prices are defined as the marginal change of the objective function value driven by a marginal change in one restriction, assuming that all other parameters remain unchanged (Baker 2000, McCarl and Spreen 2008).

One example: It can be seen in Table E4 that Japan buys coal at US \$61/ton. We know from Table E3 that Japan buys all coal from Australia. But we also know from Table E2 that Australia's CIF costs total only US \$60/ton. So why does Japan pay US \$1/ton more? Let's assume Japan would demand one more unit (in the model's case one million tons of steam coal), then Indonesia would supply this one unit as the cheapest supplier for Japan (remember, Australia is sold out). In fact, this would initially save the market US \$14/ton (US \$60/ton Australian CIF costs to Japan minus US \$46/ton Indonesian CIF costs to Japan). South Africa would then deliver 1 unit to India, the missing unit from Indonesia, which would cost the market US \$11/ton (US \$58/ton South African CIF costs to India minus US \$47/ton Indonesian CIF costs to India). The unit that South Africa is supplying to India is

Table E4 WorldCoal 2006 market prices versus real 2006 market prices

WorlCoal market prices (US $/ton)	EU	JA	KR	TW	CH_IM	IN	LA	US_IM	AS_OT
WorldCoal CIF results:	69	61	60	59	59	61	64	63	60
Actual 2006 CIF market prices:	64	65	56	56	56	56	–	–	56
Difference (%):	0.0	0.0	0.0	0.0	0.0	0.0	–	–	0.0

Sources: WorldCoal; Endres (2008); Real CIF Price calculations based on average Argus/McCloskey Index Report data from 2006 reports.

now missing in Europe; thus Russia, as the global marginal supplier, would deliver 1 more unit to Europe, which costs the market US $4/ton (US $69/ton Russian CIF costs to Europe minus US $65/ton South African CIF costs to Europe). In total, the 1 additional unit that Japan demands costs the market US $1/ton (–14 + 11 + 4 = 1). That is why GAMS correctly calculates the market price for Japan as US $61/ton, US $1 higher than Australia's CIF costs to Japan. For a more detailed discussion about WorldCoal's ShadowPrice(sup) and ShadowPrice(q) please refer to the separate paper on this subject (Endres 2008, 28ff).

Table E4 summarizes all CIF market prices calculated by GAMS. WorldCoal's results are compared to the average real 2006 CIF prices. WorldCoal comes very close to the real average market price levels. The model's CIF prices differ on average only 6% from the real 2006 CIF prices. Thus, we can state that the nonlinear WorldCoal model has been successful in approximating the 2006 world coal market.

The reasons on why the prices differ (on average, the real average world market prices for 2006 are slightly below WorldCoal's) are various. (1) WorldCoal takes the average of 2006 FOB marginal cost data, but prices started to show above-average volatility and continuously moved up during the year. (2) The same is true for freight rates, which also showed high volatility. In addition, it was assumed that freight rates are calculated based on a simple linear function dependent on distance d. In reality, each freight route is a market on its own resulting in prices that can differ from this approximation. (3) WorldCoal is a simplified approximation of the coal market, not taking into account (a) coal qualities, (b) power plants' variations, (c) each individual consumer and producer, and (d) strategic long-term pricing. For a more detailed discussion about reasons for the difference between WorldCoal's results and the real market, please also refer to the separate paper on this subject (Endres 2008, 35ff).

Results for the Scenario 2015

The reference case 2006 discussed in Section 4.5 and Appendix E shows the methodology with which WorldCoal can be used to analyze a certain scenario. Today, in 2008 when this study was written, we already know that the global coal trade is changing. The increased demand and resulting increase in export mine capacity utilization (please refer to Fig. 5.11) have resulted in coal price levels during the summer of 2008 that no one could ever have imagined.

Scenario 2015 Assumptions

IEA Energy Outlook 2007 forecasts that worldwide energy demand will increase by about 55% between 2005 and 2030. During the same time period IEA predicts that coal consumption will increase by 70% or 2.2% per annum. As discussed, this is largely driven by China and India. In the 2015 scenario I conservatively assume coal trade compound annual growth rate of 3% between 2006 and 2015, resulting in the trade volume prediction of 775 million tons in 2015.

$$\mathrm{CAGR}(t_0, t) = \left(\frac{q(t)}{q(t_0)}\right)^{\frac{1}{n}} - 1 \tag{E14}$$

$$q(2015) = q(2006)\,[\mathrm{CAGR}(2006{,}2015) + 1]^n \approx 775 \text{ million tons.} \tag{E15}$$

775 million tons of trade volume in 2015 equals a 30%, or 180 million ton, increase over the 2006 volume. The following simplifying assumptions were made about the **supply distribution** of the additional 180 million tons for the 2015 WorldCoal scenario:

- China will cease exporting.
- South Africa will not be able to increase exports beyond 2006 because of infrastructure constraints.
- Indonesia will only be able to export 10% more, driven by an increase in domestic coal demand.
- The United States can export at the most 10% more than in 2006.
- The three remaining supply regions Australia, Colombia, and Russia will supply the lacking volume. For simplicity the additional volume to these three regions was allocated based on the weighted export volume share of 2006.

The **demand distribution** is adjusted in such a way that coal import demand in all demand regions other than India and China is assumed to increase by 15%. The remaining volume is allocated to China and India resulting in 2015 coal demand import volumes for these two countries that are a multiple of 2006 volumes.

To get a reasonable result for our 2015 scenario a 3% increase (r) for all costs was assumed. Therefore, the objective function changes as follows:

$$\min \sum_i \sum_j (\mathrm{TC}_{ij}) q_{ij} (1 + r)\, n \tag{E16}$$

The freight cost function CTS_{ij} and the FOB supply functions CPR_i were left unchanged. Thus, all restrictions for the 2015 WorldCoal scenario have been described.

Scenario 2015 Results

The 2015 scenario also requires that the demand is met by minimizing total costs to the entire market. Again, supplied volumes are largely concentrated around one or two suppliers per demand region. As can be seen in Fig. E4 because China no longer exports and because Indonesia is restricted in terms of its growth, Russia takes a bigger share in Asia but remains the marginal and thus most expensive supplier for the entire world. South Africa now delivers virtually all of its exports to India.

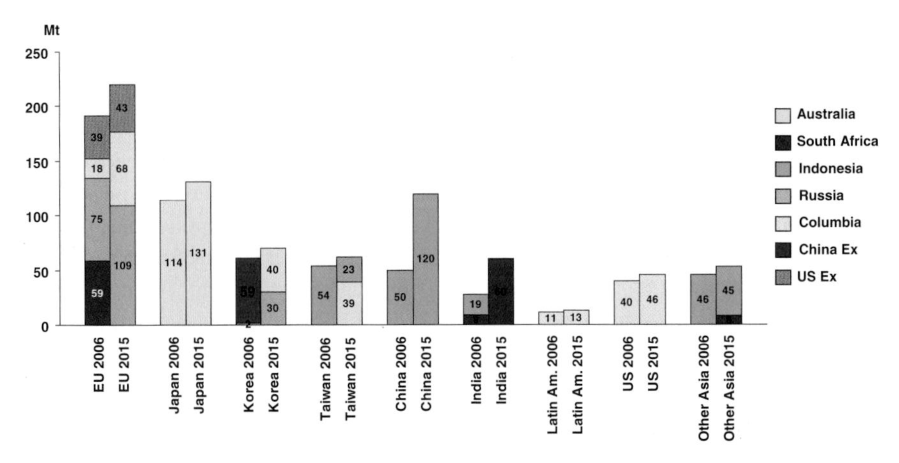

Fig. E4 WorldCoal's quantity distribution 2006 and 2015 (Source: WorldCoal; Endres 2008, Author's analysis)

The forecast market prices reach a level that is up to three times higher than that in 2006. This market price increase was expected because of increased demand and increased costs. Please refer to Fig. E5 for the price distribution 2006 vs. 2015.

Implications

WorldCoal was programmed in GAMS to model the international seaborne steam coal trade market. The planning phase resulted in a nonlinear system of equations. With simplifying model restrictions, WorldCoal then calculates the optimal market equilibrium given the total cost–minimizing objective function. The model's results for the reference year 2006 were surprisingly close to the real market. Market prices differed on average only by about 6%, despite the many simplifications, including but not limited to the one-quality assumption. This is an indication that the 2006 coal

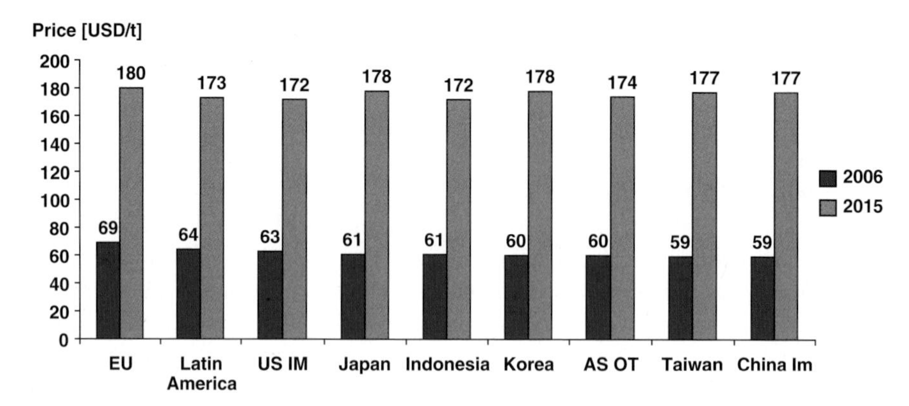

Fig. E5 WorldCoal's CIF prices 2006 and 2015 (Source: WorldCoal; Endres 2008, Author's analysis)

market operated close to the theoretical market equilibrium in a perfect competitive world.

Shadow prices were introduced for sensitivity analysis. ShadowPrice(*q*) identifies which supplier will fill a quantity that another supplier may not be able to fulfill. ShadowPrice(sup) identifies the market power that one supplier has, always within the defined restrictions. ShadowPrice(dem) signifies the market price or the marginal opportunity cost for the marginal supplier to that region. The model was also run replacing the minimizing total cost (COST) objective function with a maximizing total profit objective function. Except for a small market adjustment that reduced the quantity supplied to India by 16 million tons everything else remained constant. This is also an indication that the market seemed to be close to its optimum and there was little, if any, strategic supply shorting by the market's supplier.

The 2015 scenario restrictions were such that the supply becomes even more concentrated, with many demand regions having only one supplier. In the real coal world, this will not happen. It is unlikely that one demand region will take the strategic risk of depending on one single supply region for any longer period of time, even if it costs money. Such strategic thoughts could not be modeled into WorldCoal, however.

The 2015 scenario forecasted prices in the range of US $170–180/ton. These prices are up to three times higher than the 2006 price levels. After WorldCoal programming and analysis was finished in 2007, coal prices already skyrocketed to above US $220/ton in July 2008, driven by Asia. Thus, WorldCoal at least forecast the trend, also driven by Asia. The question of what will happen from here still remains, which I discussed in Chapter 7.

However, WorldCoal only models an entire year. Thus, any intra-year volatility cannot be described, and in fact it wasn't meant to be. When modeling a global market, the point is to get an understanding of what will happen when certain scenarios are realized. They are also useful for predicting long-term market trends. As such, I believe that WorldCoal is fulfilling its purpose for the global coal trade market.

However, I am aware that WorldCoal has many limitations that future researchers will hopefully reduce. But as with any model, it has to remain manageable and thus will always require simplifications.

Finally, WorldCoal analyzed the coal market under the assumption of perfect competition. However, I have shown in this study that coal supply has oligopolistic tendencies. Thus, it will be interesting to compare the WorldCoal model to a Cournot model. While building a Cournot model is not a subject of this study, I will try to build a basis for such model by studying Cournot. The goal of Chapter 5 (see also Appendix G) is to calculate the Cournot equilibrium with increasing marginal cost – as is the case for the coal market and any other raw materials supply market – rather than constant marginal cost.

Appendix F
GAMS Programming Syntax of WorldCoal

```
*WorldCoal 2006
*--------------------------------------
Sets
i suppliers   /au
               sa
               id
               ru
               co
               ch_ex
               us_ex/
j consumers   /eu
               ja
               kr
               tw
               ch_im
               in
               la
               us_im
               as_ot /;
*--------------------------------------
Parameters
***Supply:
q_sup_min_1(i)    minimum supply in i
                  in mtpy in section 1      /au     0
                                             sa     0
                                             id     0
                                             ru     0
                                             co     0
                                             ch_ex  0
                                             us_ex  0/
q_sup_max_1(i)    maximum supply in i
                  in mtpy in section 1      /au     83
```

```
                                                    sa      54
                                                    id      124
                                                    ru      61
                                                    co      51
                                                    ch_ex   21
                                                    us_ex   20/
q_sup_min_2(i)      minimum supply in i
                    in mtpy in section 2           /au      0
                                                    sa      0
                                                    id      0
                                                    ru      0
                                                    co      0
                                                    ch_ex   0
                                                    us_ex   0/
q_sup_max_2(i)      maximum supply in i
                    in mtpy in section 2            /au      31
                                                     sa      14
                                                     id      47
                                                     ru      14
                                                     co      18
                                                     ch_ex   38
                                                     us_ex   19/
***Demand:
q_dem(j)            demand in market j
                    in mtpy                         /eu      191
                                                     ja      114
                                                     kr      61
                                                     tw      54
                                                     ch_im   50
                                                     in      28
                                                     la      11
                                                     us_im   40
                                                     as_ot   46/
***Function Parameters:
a_1(i)             axis intercept for section 1
                   /au 16, sa 28, id 19, ru 35, co 23,
                   ch_ex 33, us_ex 30/
b_1(i)             slope of production cost function for
                   section 1
                   /au 0.09, sa 0, id 0.01, ru 0.17, co 0,
                   ch_ex 0, us_ex 0/
a_2(i)             axis intercept for section 2
                   /au 24, sa 28, id 21, ru 46, co 23,
                   ch_ex 33, us_ex 30/
```

```
b_2(i)              slope of production cost function
                      for section 2
                    /au 0.67, sa 1.17, id 0.28, ru 1.07,
                    co 1.2, ch_ex 0.27, us_ex 0.75/
*-------------------------------------
Table d(i,j)        sea distance in nautical miles

        eu      ja    kr     tw     ch_im   in    la    us_im   as_ot
au      11644   4272  4440   4031   4216    6423  6923  9136    4008
sa      7025    7586  7528   6338   6196    3823  8734  8082    4823
id      9041    2848  2630   1770   1791    3482  8785  11585   732
ru      974     772   471    1404   1624    5735  6311  5327    2994
co      4389    8237  8655   9534   9801    8758  1951  1487    11084
ch_ex   10698   1074  520    824    100000  5136  7224  10039   2395
us_ex   3969    9165  9583   10462  10729   8677  2879  100000  10782

Parameter c_ts(i,j)  transshipment cost in dollar per ton;
                     c_ts(i,j) = 6.5 + 0.002*d(i,j);
*-------------------------------------
Variable q_s_1(i)     quantity supplied by i per year in section 1;
Variable q_s_2(i)     quantity supplied by i per year in section 2;
Variable q_1(i,j)     quantity shipped from i to j per year of
                        section 1;
Variable q_2(i,j)     quantity shipped from i to j per year of
                        section 2;
Variable q_tot(i,j)   total quantity shipped from i to j per year;
Variable c_prod_1(i)  production cost function in section 1;
Variable c_prod_2(i)  production cost function in section 2;
Variable c(i)         cost for each supplier;
Variable c_tot        total cost;
Positive Variable q_s_1;
Positive Variable q_s_2;
Positive Variable q_1;
Positive Variable q_2;
Positive Variable q_tot;
Positive Variable c_prod_1;
Positive Variable c_prod_2;
Positive Variable c;
*-------------------------------------
Equations
        totalcost           objective function
        cost(i)             cost for each supplier
        prod_cost_1(i)      production cost function in section 1
        prod_cost_2(i)      production cost function in section 2
        sup_max_1(i)        max supply section 1
```

```
        sup_max_2(i)         max supply section 2
        dem(j)                    demand
        sup1(i)                   real supply in section 1
        sup2(i)                   real supply in section 2
        suptot(i,j)               total supply;
prod_cost_1(i)  .. c_prod_1(i) =e= (a_1(i) + b_1(i)* q_s_1(i));
prod_cost_2(i)  .. c_prod_2(i) =e= (a_2(i) + b_2(i)* q_s_2(i));
sup1(i)         .. sum(j, q_1(i,j)) =e= q_s_1(i);
sup2(i)         .. sum(j, q_2(i,j)) =e= q_s_2(i);
suptot(i,j)     .. q_tot(i,j) =e= q_1(i,j) + q_2(i,j);
cost(i)         .. c(i) =e= sum(j, c_ts(i,j)*q_tot(i,j)
                           + c_prod_1(i)*q_1(i,j)
                           - 0.5*(c_prod_1(i) - a_1(i))*q_1(i,j)
                           + c_prod_2(i)*q_2(i,j)
                           - 0.5*(c_prod_2(i)-c_prod_1(i))*q_2(i,j));
totalcost      .. c_tot =e= sum(i, c(i));
sup_max_1(i)   .. q_s_1(i) =l= q_sup_max_1(i);
sup_max_2(i)   .. q_s_2(i) =l= q_sup_max_2(i);
dem(j)         .. sum(i, q_tot(i,j)) =e= q_dem(j);

Model WorldCoal /all/;
Solve WorldCoal using nlp minimizing c_tot;
Parameter  fob(i)  free on board cost;
            fob(i) = c_prod_2.l(i);
Parameter  cif(i,j)  cif delivery;
            cif(i,j) = (fob(i) + c_ts(i,j));
Parameter  q_s_tot(i)  total supply of each supplier;
            q_s_tot(i) = sum(j, q_tot.l(i,j));
```

Appendix G
Industrial Structure: Game Theory and Cournot

In this analysis I aim to illustrate Cournot competition by relaxing the standard assumption that marginal costs are constant. As with most raw material production markets, it has been empirically shown and discussed in Section 5.4 that marginal steam coal FOB costs for steam coal are increasing exponentially rather than remaining constant. The discussion about Cournot with increasing marginal cost probably started in the 1950s; thus, it is not novel in itself. However, the literature is generally of a very complex nature and not easy to find. Standard text books, such as Tirole (1988), do mention the possibility of nonlinear cost functions in an abstract form, but examples are for linear cost functions. Some older and newer literature, such as Fisher (1961) and Kopel (1996), discusses nonlinear cost functions in more detail but does not mention the impact of price and profits, as I will do here.

Cournot competition is an economic game theory model used to describe industry structure. It was named after Antoine Augustin Cournot (1801–1877) after he observed competition in a spring water duopoly. Cournot's model has the following features:

- There is more than one firm and all firms produce a homogeneous product;
- Firms do not cooperate;
- Firms have market power;
- The number of firms is fixed;
- Firms compete in quantities and choose quantities simultaneously;
- There is strategic behavior by the firms.

An essential assumption of this model is that each firm aims to maximize profits. The following is also assumed:

- Price is a commonly known decreasing function of total output;
- All firms know n, the total number of firms in the market;
- Each firm has a cost function $c_i(q_i)$;
- The cost functions may be the same or different among firms;
- The market price is set at such a level that demand equals the total quantity produced by all firms.

Cournot has the following key implications: (1) output is greater with Cournot duopoly than monopoly (in extreme cases it would be the same), but lower than perfect competition. (2) Price is lower with Cournot duopoly than monopoly (in extreme cases it would be the same), but not as low as with perfect competition.

According to this model, the firms have an incentive to form a cartel, effectively turning the Cournot model into a monopoly. However, cartels are usually illegal, so firms have some motive to tacitly collude using self-imposing strategies to reduce output, which (ceteris paribus) raises the price and thus increases profits, if the price effect is bigger than the output effect.

The study of Cournot and thus game theory is essential for the coal market as it approximates how producers can influence the market price and market quantity. The coal market is not a perfect competition and it is not a monopoly. As such the Cournot duopoly provides insight from a theoretical perspective into how the coal market behaves in the practical world and how much market power even smaller players have.

For this analysis, the following variables and abbreviations are defined:

D	Demand
$p(Q)$	Inverse demand function
$c_i(q_i)$	Cost function of firm i
n	Total number of firms competing
Π	Profit
Q	Total quantity produced in industry
Q_S	Fictitious market volume (i.e., in perfect competition with constant c_v)
q_i	Quantity produced by company i
q_{-i}	Quantity produced by all other companies (except company i)
p	Price
$C(q) = \text{TC}$	Total cost (= fixed cost + variable cost x quantity)
ATC	Average total cost
$C_f = \text{FC}$	Fixed cost
$c_v = \text{VC}$	Variable cost
$[x]^*$	x in the equilibrium, therefore q^* equals the quantity in the equilibrium
AVC	Average variable cost
$C'(q) = \text{MC}$	Marginal cost, the derivative of total cost equals marginal cost
TR	Total revenue
(TR)' = MR	Marginal revenue, the derivative of total revenue equals marginal revenue
FOC	First order condition (does not equal the derivative, but a condition that uses the derivative)
PP	Price Premium
ΠP	Profit Premium

Before discussing game theory and Cournot, two reference cases should be discussed and understood: case 1 perfect competition and case 2 monopoly. The following discussion is useful to get a better understanding of Cournot in general.

Reference Case 1: Perfect Competition

Perfect competition is characterized by the fact that each market player has no market power and cannot influence the price. The market price cannot be influenced because the total quantity can only be minimally influenced by any individual market player. We already briefly introduced perfect competition when discussing price theory in Section 5.3.4.

It is important to study perfect competition for the coal market as well. The key principal – that the market price equals the marginal cost of the marginal supplier – also approximates the real coal market. When discussing Cournot, perfect competition is a reference case and forms the lower boundary for the price in Cournot where n is indefinite.

The profit of company i is calculated by taking the total quantity produced by the company times price and deducting the variable cost times quantity as well as all fixed costs, see formula (G1).

$$\Pi_i = q_i(p - \mathrm{ATC}_i) = pq_i - C(q_i) = \mathrm{TR}_i - \mathrm{TC}_i = pq_i - c_v q_i - C_f, i = 1, 2, \ldots n \quad \text{(G1)}$$

- I omit i for simplification in the following formulae.
- The profit has to be 0 or positive in order for any quantity to be produced.
- In this case the VC itself does not depend on the quantity, as such the VC is assumed to be constant.

Optimality conditions for the correct choice of quantity:

- $\mathrm{MR} - \mathrm{MC} = 0 \Leftrightarrow \mathrm{MR} = \mathrm{MC}$
- $\mathrm{TR} \geq \mathrm{TC} \Rightarrow P \geq \mathrm{AVC}$ (otherwise you wouldn't produce and set $q = 0$)

Implications: the price in perfect competition in social welfare optimum is set in such a way that, $p = \mathrm{MR} = \mathrm{MC} \Rightarrow \Pi_i = 0 \Rightarrow$, where no dead weight loss occurs, which is shown below.

$$\frac{\partial \Pi}{\partial q} = \frac{\partial p}{\partial q} q + p \frac{\partial q}{\partial q} - \frac{\partial C(q)}{\partial q} = 0; \text{is equivalent to FOC of } \Pi = pq - C(q) \quad \text{(G2)}$$

- $\frac{\partial p}{\partial q} q = 0$ because the price does not depend on the quantity.
- $p \frac{\partial q}{\partial q} = p$ because the derivative of q equals one in the quantity competition.

$$\Rightarrow p - \frac{\partial C(q)}{\partial q} = 0 \Leftrightarrow p = \frac{\partial C(q)}{\partial q} = C'(q) \Rightarrow \text{MR} = \text{MC} \tag{G3}$$

$$p = \frac{\partial C(q)}{\partial q} = \frac{\partial c_v}{\partial q} q + c_v \frac{\partial q}{\partial q} + \frac{\partial C_f}{\partial q} = c_v = \text{MC} \tag{G4}$$

- $\frac{\partial c_v}{\partial q} = 0$ because the variable costs do not depend on the quantity.
- $\frac{\partial C_f}{\partial q} = 0$ because the variable costs do not depend on the quantity.

Figure G1 shows graphically perfect competition and the fact that the market price is determined by the marginal cost of the marginal supplier to the market.

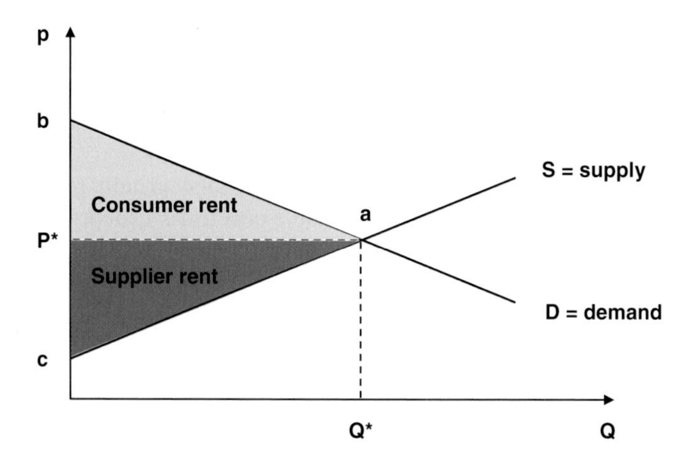

Fig. G1 Graphical figure of reference case perfect competition (Source: Author)

Reference Case 2: Monopoly

In a monopoly there is only one player, in our case one coal producer, that has full market power and full control of the market price by setting its quantity. In perfect competition the price does not depend on the quantity of one market player, in monopoly naturally the price depends on the monopolist's supplied quantity.

The monopoly is at the opposite extreme to perfect competition. Here one player has all the market power. This is important to study also for the coal market, as it illustrates the maximum producer (or supplier) rent that can be generated. Cournot results will be somewhere in between monopoly and perfect competition results.

Monopoly Calculations

In monopoly the price is a function of the quantity supplied. Therefore the profit function for a monopoly changes slightly from that of perfect competition.

$$\Pi_i = p(q_i)q_i - C(q_i) = \text{TR}(q_i) - \text{TC}(q_i) = p(q_i)q_i - c_v q_i - C_f, i = 1 \quad \text{(G5)}$$

- I omit i because there is only one market player, the monopolist, in the following formulae.
- From here onward I also substitute q_i with Q because the monopolist supplies the entire market quantity.
- The profit has to be 0 or positive in order for any quantity to be produced.
- In this case the VC itself does not depend on the quantity, as such the VC is assumed to be constant.

Optimality conditions for the correct choice of quantity:

- $\text{MR} - \text{MC} = 0 \Leftrightarrow \text{MR} = \text{MC}$
- $TR \geq TC$

Implications: $\text{MR} = \text{MC} \,\&\, p \geq \text{MC} \Rightarrow \Pi \geq 0 \Rightarrow$ social welfare optimum is only reached if the monopolist is able to perfectly discriminate for each consumer, which requires the monopolist to know each consumer's ability and willingness to pay for a product (thus know each consumer's utility). Because the price depends on the quantity, the MR does not equal p anymore (as in perfect competition) but it equals $p'(Q)Q + p(Q)$.

$$\frac{\partial \Pi}{\partial Q} = \frac{\partial p(Q)}{\partial Q}Q + p(Q)\frac{\partial Q}{\partial Q} - \frac{\partial C(Q)}{\partial Q} = 0; \quad \text{(G6)}$$

is equivalent to FOC of $\Pi = p(Q)Q - C(Q)$

$$p(Q)\frac{\partial Q}{\partial Q} = p(Q); \text{because the derivative of Q equals one} \quad \text{(G7)}$$

$$\begin{aligned} &\Rightarrow \frac{\partial p(Q)}{\partial Q}Q + p(Q) - \frac{\partial C(Q)}{\partial Q} = p'(Q)Q + p(Q) - C'(Q) = 0 \\ &\Leftrightarrow p'(Q)Q + p(Q) = C'(Q) \Rightarrow \text{MR} = \text{MC} \end{aligned} \quad \text{(G8)}$$

Please note that $p'(Q)$ is negative since $p(Q)$ is decreasing function (for instance, $p = a - bQ^x$), or in other words price and quantity are negatively correlated. The consumers pay a premium of $-p'(Q)Q$ to the monopolist compared to price in perfect competition:

$$\Rightarrow p'(Q)Q + p(Q) = \frac{\partial C(Q)}{\partial Q} = \frac{\partial c_v}{\partial Q}Q + c_v\frac{\partial Q}{\partial Q} + \frac{\partial C_f}{\partial Q} = c_v \quad \text{(G9)}$$

- $\frac{\partial c_v}{\partial q} = 0$ because the variable costs do not depend on the quantity.
- $\frac{\partial C_f}{\partial q} = 0$ because the fixed costs do not depend on the quantity.

Figure G2 shows graphically monopoly and the fact that the market price is no longer determined by the marginal cost. There is a premium charged to the consumers. The cost of monopoly is called the dead weight loss. This dead weight loss is the economical reason why an economy should usually try to avoid monopolies.

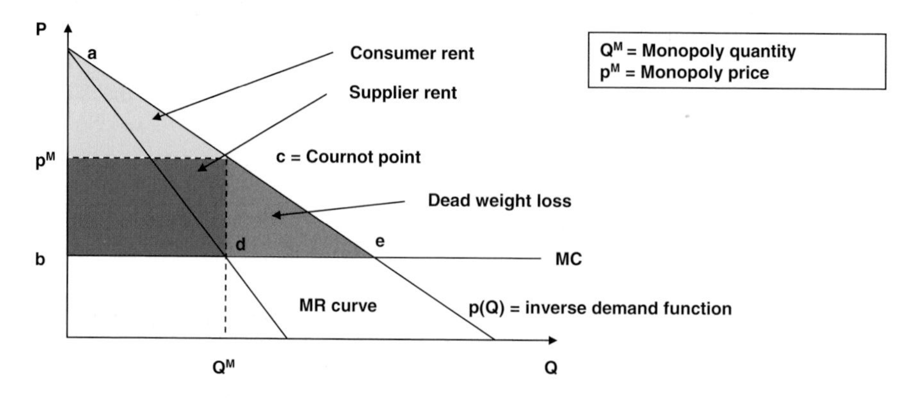

Fig. G2 Graphical figure of reference case monopoly (Source: Author)

Lerner Index

The Lerner index in a monopoly describes that part of the extra profit a monopolist can generate compared to perfect competition. As discussed previously, market power here means that the monopolist can influence the price $p(Q)$with his own quantity.

$$\frac{\partial \Pi}{\partial Q} = \frac{\partial p(Q)}{\partial Q} Q + p(Q) \frac{\partial Q}{\partial Q} - \frac{\partial C(Q)}{\partial Q} = 0; \tag{G10}$$

is equivalent to FOC of $\Pi = p(Q)Q - C(Q)$

$$\Leftrightarrow \frac{\partial p(Q)}{\partial Q} Q + p(Q) = \frac{\partial C(Q)}{\partial Q} = \mathrm{MC} \tag{G11}$$

$$\frac{\partial p(Q)}{\partial Q} = p'(Q) \tag{G12}$$

$$\Leftrightarrow p(Q) - \mathrm{MC} = -p'(Q)Q \,|: p(Q) \tag{G13}$$

$$\Rightarrow \frac{p(Q) - MC}{p(Q)} = -\frac{p'(Q)Q}{p(Q)} = \frac{1}{|\eta|} \tag{G14}$$

$p(Q) - \mathrm{MC}$, which in a monopoly is positive or at least 0, equals the monopoly markup. The Lerner index is defined as this mark-up relative to the price (or how the markup moves with the market price):

$$\eta := \frac{\partial Q}{\partial p(Q)}\frac{p(Q)}{Q} = Q'(p)\frac{p(Q)}{Q} \Rightarrow \frac{1}{\eta} = \frac{\partial p(Q)}{\partial Q}\frac{Q}{p(Q)} = p'(Q)\frac{Q}{p(Q)} = \frac{p'(Q)Q}{p(Q)} \quad \text{(G15)}$$

η(eta) is the price elasticity of the demand as defined above. This elasticity causes the monopoly market power

- $\eta < -1$ means elastic demand.
- $\eta = -1$ unit elastic demand (Einheitselastisch), for more details please refer to Uni Basel 2007.
- $\eta > -1$ means inelastic demand.

$\frac{1}{|\eta|}$ equals the Lerner index (the absolute value of the inverse of the elasticity η)

$$\Rightarrow \text{MR} = p'(Q)Q + p(Q) = p(Q)\left(\frac{p'(Q)Q}{p(Q)} + \frac{p(Q)}{p(Q)}\right)$$
$$= p(Q)\left(\frac{p'(Q)Q}{p(Q)} + 1\right) = p(Q)\left(\frac{1}{\eta} + 1\right) \quad \text{(G16)}$$

$$\Rightarrow p(Q)\left(\frac{1}{\eta} + 1\right) = \text{MC, because MR equals MC} \quad \text{(G17)}$$

***Implications*:** The more inelastic the demand, the higher the monopoly markup or premium; however, the markup alone does not indicate the monopolist's profit. The monopolist's profit is the highest when the demand is unit elastic (as above); therefore the monopolist would always try to avoid a price with elastic or inelastic demand.

A Monopoly with Linear Demand Function

Here I will briefly demonstrate why the slope of MR is twice the slope of the demand function in a monopoly with linear demand as shown in Fig. G2 above.

I set $p(Q) = a - bQ$, the inverse demand function

$$\Pi = (a - bQ)Q - C(Q) \quad \text{(G18)}$$

$$\frac{\partial \Pi}{\partial Q} = (a - bQ) + (-b)Q - C'(Q) = a - 2bQ - \text{MC} = 0 \quad \text{(G19)}$$

$$\Leftrightarrow \text{MC} = a - 2bQ = \text{MR} \quad \text{(G20)}$$

Cournot Competition

When studying Cournot, which means studying simultaneous quantity competition, one also should study Bertrand, which means studying simultaneous price competition, and Stackelberg, which means studying dynamic quantity competition. All three are relevant to the coal market:

- Cournot – simultaneous quantity competition: smaller players in the coal market take the market price as given, but here we study how they still have some market power by adjusting their output quantity.
- Bertrand – simultaneous price competition: in some coal markets the output quantity may be given or fixed due to logistics or legislation; here I study how the market players have market power by adjusting their own price to the MC of the marginal producer for that market.
- Stackelberg – dynamic quantity competition: here I study how a new market entrant affects a monopolist's position. This is important for the coal market as well, as monopolists may be challenged by new entrants in smaller submarkets of the coal industry.

I start with the simultaneous quantity competition: the classic Cournot model. As discussed previously, the Cournot model allows studying competitive situations that are different from perfect competition and the monopoly. When looking for the Nash equilibrium in the Cournot game n symmetrical companies that compete on quantity are studied.

The case is different from perfect competition as there are only n symmetrical firms and not an unlimited number of firms; it is different from monopoly as there is more than one firm. As such, it is somewhere in-between and in fact describes a market such as the coal market more realistically.

In Cournot, all companies act simultaneously and maximize their profit by adjusting their own supply quantity. As such, they have to react to every possible quantity of the other players; they therefore build a reaction function for it.

Calculating Each Player's Equilibrium Quantity q_i*

$$\Pi_i = (a - bQ)q_i - C(q_i) = \left(a - b\sum\nolimits_{i=1}^{n} q_i\right) q_i - C_f - c_v q_i \tag{G21}$$

$$C(q_i) = C_f + c_v q_i \tag{G22}$$

$$p(Q) = p(q_1, q_2, ..., q_n) = a - bQ = a - b\sum\nolimits_{i=1}^{n} q_i, \text{ inverse demand function} \tag{G23}$$

It is assumed that all companies have symmetrical cost functions, if not the cost function would be $C_i(q_i) = C_{f_i} + c_{v_i} q_i$. The fictitious market volume is defined as $Q_S := \frac{a-c_v}{b}$ where $p = c_v$ (perfect competition).

Optimality conditions for the correct choice of quantity:

- $\mathrm{MR} - \mathrm{MC} = 0 \Leftrightarrow \mathrm{MR} = \mathrm{MC}$
- $\mathrm{TR} \geq \mathrm{TC}$

$$\Rightarrow \frac{\partial \Pi_i}{\partial q_i} = (a - bQ) + (-b)q_i - c_v = \left(a - b\left(q_i + \sum\nolimits_{j \neq i} q_j\right)\right) - bq_i - c_v$$
$$= a - 2bq_i - b\sum\nolimits_{j \neq i} q_j - c_v = 0 \tag{G24}$$

- $Q = q_i + \sum_{j \neq i} q_j$
- Note: This step is required to later be able to build the reaction function where we need to separate all q_i

Next I build the reaction function by resolving for q_i for every possible quantity of the remaining companies q_{-i}

$$\Rightarrow R_i(q_{-i}) = q_i = \frac{a - b\sum_{j \neq i} q_j - c_v}{2b} \tag{G25}$$

The reaction functions for all companies are the same as above because all companies are symmetrical. Therefore, one can set q_i equal to q in the equilibrium and look for the appropriate quantity of each market player where n equals the number of all market players.

$$\Rightarrow q = \frac{a - b(n-1)q - c_v}{2b} \Leftrightarrow q2b + b(n-1)q = a - c_v \tag{G26}$$

$$\Leftrightarrow qb(2 + n - 1) = qb(1 + n) = a - c_v \tag{G27}$$

$$\Leftrightarrow q^* = \frac{a - c_v}{b(n+1)} \tag{G28}$$

Thus, the equilibrium quantity q_i^* of each market player is found to be as above in the Cournot quantity competition with two players; if company i increases its quantity, then company j should decrease it, and vice versa.

For comparison I calculate with $Q_S := \frac{a-c_v}{b}$; using (G28) this translates to $q^* = \frac{Q_S}{n+1}$. For example, with two players, and if player 1 does not know the quantity that the other player definitely offers, player 1 should offer one-third of the total demanded quantity that would be offered in perfect competition.

Calculating the Total Market Quantity Q^* in Equilibrium

$$Q^* = nq^* = n\frac{a - c_v}{b(n+1)} \tag{G29}$$

For comparison I calculate with $Q_S := \frac{a-c_v}{b}$; this translates to

$$Q^* = n\frac{Q_S}{n+1} = \frac{n}{n+1}Q_s \tag{G30}$$

Calculating the Equilibrium Market Price p^*

Substituting Q using the formula (G29) above the equilibrium price is calculated as follows:

$$\begin{aligned} p^*(Q) &= a - bQ = a - b\frac{n(a-c_v)}{b(n+1)} = a - \frac{na - nc_v}{(n+1)} \\ &= \frac{an + a - na + nc_v}{n+1} = \frac{a + nc_v}{n+1} - c_v + c_v = \frac{a + nc_v - c_v n - c_v}{n+1} + c_v \end{aligned} \tag{G31}$$

$$\Rightarrow p^*(Q) = \frac{a - c_v}{n+1} + c_v \tag{G32}$$

(Above the mathematical ploy to add and deduct c_v was used.)

For comparison I calculate with $Q_S := \frac{a-c_v}{b}$, this translates to $p^*(Q) = \frac{bQ_S}{n+1} + c_v$.

If one now lets n increase without limitation (to approach perfect competition), then the following applies, which compares to perfect competition:

$$p^*(Q) = \frac{a - c_v}{n+1} + c_v \xrightarrow{\lim n \to \infty} c_v = p^*(Q) \tag{G33}$$

Thus, the more the players, the closer one gets to perfect competition.

Calculating the Equilibrium Profit Π^* for Each Market Player

$$\Pi_i = pq_i - C(q_i) = (p - \mathrm{MC}_i)q_i - C_f \tag{G34}$$

The quantity of each player $q_i^* = \frac{a-c_v}{b(n+1)}$ was defined in formula (G28) above, and the equilibrium market price $p^* = \frac{a-c_v}{n+1} + c_v$ was defined in formula (G32). Below I will use c_v for MC_i:

$$\Pi_i^*(n) = (p^* - \mathrm{MC}_i)q_i^* - C_f = \left(\frac{a - c_v}{n+1} + c_v - c_v\right)\left(\frac{a - c_v}{b(n+1)}\right) - C_f$$
$$= \frac{1}{b}\left(\frac{a - c_v}{n+1}\right)^2 - C_f \qquad \text{(G35)}$$

Calculating How Many Companies *n* Will Enter the Market

As discussed previously, in the Cournot Game the companies decide to enter the market before the competition begins. Thus, they calculate the equilibrium and then decide whether they will join the market or not. The company entering the market does so without any regulation; thus n^{FE} where FE means Free Entry:

$$\Pi_i^*(n) = \frac{1}{b}\frac{(a - c_v)^2}{(n+1)^2} - C_f = 0 \qquad \text{(G36)}$$

$$\Leftrightarrow \frac{1}{b}\frac{(a - c_v)^2}{(n+1)^2} = C_f \qquad \text{(G37)}$$

$$\Leftrightarrow \frac{1}{b}\frac{(a - c_v)^2}{C_f} = (n+1)^2 \qquad \text{(G38)}$$

$$\Leftrightarrow \left(\frac{1}{b}\frac{(a - c_v)^2}{C_f}\right)^{\frac{1}{2}} - 1 = n^{FE} \qquad \text{(G39)}$$

Summary Comparing Cournot to Monopoly

The formulas above also work in the monopoly reference case where n equals 1. Below I compare all formulas for monopoly and Cournot:

	Monopoly	Cournot
q^*	$\frac{a - c_v}{2b}$	$\frac{a - c_v}{b(n+1)}$
Q^*	$\frac{a - c_v}{2b}$	$n\frac{a - c_v}{b(n+1)}$
p^*	$\frac{a - c_v}{2} + c_v$	$\frac{a - c_v}{n+1} + c_v$
Π^*	$\frac{1}{b}\left(\frac{a - c_v}{2}\right)^2 - C_f$	$\frac{1}{b}\left(\frac{a - c_v}{n+1}\right)^2 - C_f$

Bertrand and Stackelberg

After having studied Cournot, simultaneous quantity competition with constant marginal cost, I now turn to Bertrand, simultaneous price competition, and Stackelberg, dynamic quantity competition.

Bertrand, Simultaneous Price Competition

In simultaneous price competition, the quantity is given and the market players compete on price for that given quantity. The quantity may be given or fixed due to logistics or legislation. The market power that each player has by adjusting his own price to the marginal cost of the marginal producer for that market can be illustrated.

When looking for the Nash equilibrium in the Bertrand game, n symmetrical companies compete on price and not quantity.

- Here it is assumed the consumers are extremely price-sensitive, much more so than before, and act accordingly.
- Unlimited production capacity is assumed: this means that any company can produce as much as it likes.
- Thus, anyone can flood the market and thus any player has a lot of market power (poses a threat to the other players).
- The Bertrand model finds the price, which in the end logically should equal the MC of the lowest cost producer; however, in a heterogeneous game the lowest MC producers is able to increase his price to the MC of the second lowest MC producer, and as such makes a profit.
- In this model different prices p_i are calculated.

All companies act simultaneously and try to maximize their profits by adjusting their own supply price. As such, they have to react to every possible price of the other players; therefore, they build a reaction function for it.

Homogeneous Betrand Model

Homogeneous means MC_i equals $\mathrm{MC}_{j.}$ This game works with two competitors for reference purposes. The model also works with more than two (or n) companies, but not with one.

$\mathrm{MC}_i = \mathrm{MC}_j$, the quantity demanded from player i can be calculated as follows:

$$D_i(p_i,p_j) = \begin{cases} 0 & p_i > p_j \\ D(p_i)/2 & p_i = p_j \\ D(p_i) & p_i < p_j \end{cases} \Rightarrow p_i = \mathrm{MC}_i \Rightarrow \Pi_i = 0, i = 1,2 \qquad \text{(G40)}$$

Formula (G40) means that

1. if company *i*'s price is above company *j*'s price, then company *i* will not sell any product;
2. if company *i*'s price equals company *j*'s price, then company *i* will satisfy half of the market's demand; and
3. if company *i*'s price is below company *j*'s price, then company *i* will satisfy the entire market's demand at its price *p*.

Conclusion: $p_i = \text{MC}_i$ is a Nash Equilibrium, because no player has an incentive to change his price for the following reasons:

1. In the event that one player increases his price above his MC, he will immediately lose all of his consumers, thus not optimal.
2. In the event that one player was to decrease his price, he would immediately gain all consumers in the market and could increase his own share of the market demand. But he would satisfy the market by losing money because he would sell below his MC, thus also not optimal.

Heterogeneous Bertrand Model

Heterogeneous means MC_i does not equal MC_j. This game works with two competitors for reference purposes. Again, the model also works with more than two (or *n*) companies, but not with one.

Assumptions:

- $\text{MC}_i < \text{MC}_j$.
- $p_j = c_j$ and $p_i = c_j - \varepsilon$.
- Then, company *j* (even though more costly than company *i*) cannot set its price below MC_j.
- Then, company *i* has to avoid setting its price at MC_j, because then company *i* would only receive half the market demand; thus it needs to set its price just below MC_j. And this factor is set as ε.

It follows for the heterogeneous Bertrand model:

$$\Rightarrow q_i = D_i(p_i, p_j) = D(c_j - \varepsilon) \approx D(c_j) \text{ and } \Rightarrow q_j = D_j(p_i, p_j) = 0 \quad \text{(G41)}$$

$$\Rightarrow \Pi_i \geq 0, \Pi_j = 0 \quad \text{(G42)}$$

Stackelberg, Dynamic Quantity Competition

In the Cournot model the market players act simultaneously and compete on quantity. Stackelberg on the other hand studied dynamic quantity competition. Thus, here the players react to one another. The Nash Equilibrium in a quantity game is still the goal.

To analyze this case a 'Stackelberg Leader' is defined who is already in the market and a 'Stackelberg Challenger' is defined who considers entering the market. The Leader anticipates the Challenger's move and reacts in order to maximize his profit.

From a market perspective, consider the following: A monopolist, that is, a coal producer that exclusively supplies a country or a consumer, is almost always attacked by potential new entries. A monopolist (the Leader) therefore tries to create barriers to entry, or if that is not possible, then anticipates a new market entry (the Challenger) and sets his own new quantity in such a way that his profit is still maximized.

Calculations of the Stackelberg Challenger

Challenger i and Leader j are defined. The reaction function is taken and modified for two players:

$$R_i(\mathrm{q}_{-\mathrm{i}}) = q_i = \frac{a - b\sum_{j \neq i} q_j - c}{2b} \Rightarrow R_i(\mathrm{q}_\mathrm{j}) = \frac{a - b(2-1)q_j - c_i}{2b} = \frac{a - bq_j - c_i}{2b} \tag{G43}$$

***Conclusion*:** The Challenger chooses its offered quantity. This quantity will equal the quantity that he would offer in the Cournot game.

Calculations of the Stackelberg Leader

Challenger i and Leader j:

- Leader j now anticipates the move of Challenger i and has the privilege of acting first.
- Therefore, in order to calculate his optimal quantity, the Leader utilizes the reaction function of the Challenger as per (G25) above for his calculations.

$$\Rightarrow \Pi_j = (a - b(q_i + q_j))q_j - c_j q_j = \left(a - b\left(\frac{a - bq_j - c_i}{2b} + q_j\right)\right) q_j - c_j q_j$$
$$= \left(a - b\left(\frac{a + bq_j - c_i}{2b}\right)\right) q_j - c_j q_j \tag{G44}$$

One could now calculate the FOC where $\frac{\partial \Pi_j}{\partial q_j} = 0$, which I will not do here for reasons of simplicity. Next, q_j is separated from the FOC and both players' quantity, profit, and the market price are calculated.

Example for the Stackelberg Model

Instead of calculating more complex formulae, I will work with an example to illustrate how Stackelberg works.

Assumptions:

- two players 1 and 2;
- homogenous marginal cost; $c_1 = c_2 = 3$ (for simplification);
- inverse demand function; $p(Q) = 15 - Q = 15 - (q_1 + q_2)$; and
- backward induction is used.

Step 1: Calculate reaction function of Challenger (company 2)

$$\begin{aligned}
&\max_{q_2\geq 0} \Pi_2(q_1, q_2) = (15 - q_1 - q_2)q_2 - c_2 q_2 \\
&\Rightarrow \frac{\partial \Pi_2}{\partial q_2} = 15 - q_1 - 2q_2 - c_2 = 0 \\
&\Rightarrow R_2(q_1) = q_2 = \frac{15 - q_1 - c_2}{2} = \frac{12 - q_1}{2}
\end{aligned} \tag{G45}$$

Step 2: Insert the above reaction function of the Challenger into the profit function of the Leader (company 1)

$$\begin{aligned}
&\max_{q_1\geq 0} \Pi_1(q_1, q_2) = (15 - q_1 - q_2)q_1 - c_1 q_1 = \left(15 - q_1 - \frac{12 - q_1}{2}\right) q_1 - c_1 q_1 \\
&= \left(9 - q_1 + \frac{q_1}{2}\right) q_1 - c_1 q_1 \\
&\Rightarrow \frac{\partial \Pi_1}{\partial q_1} = 9 - 2q_1 + q_1 - c_1 = 0 \\
&\Leftrightarrow q_1 = 9 - c_1 = 6
\end{aligned} \tag{G46}$$

Step 3: Insert quantity of the Leader (=6) into all functions and the following set of solutions is derived:

$$\begin{aligned}
&q_2 = \frac{12 - q_1}{2} = 3 \\
&Q = 6 + 3 = 9 \\
&p(Q) = 15 - Q = 6 \\
&\Pi_1 = 6{*}6 - 3{*}6 = 18 \\
&\Pi_2 = 6{*}3 - 3{*}3 = 9 \\
&\Pi = 9 + 18 = 27
\end{aligned} \tag{G47}$$

Step 4: What would the Leader's profit be like, before the Challenger enters (i.e., the leader is still a monopolist)?

$$\Pi_1(Q) = (15 - Q)Q - c_1 Q$$
$$\Rightarrow \frac{\partial \Pi_1}{\partial Q} = 15 - 2Q - c_1 = 0$$
$$\Rightarrow Q = \frac{15 - c_1}{2} = 6 \hat{=} q_1 \tag{G48}$$
$$\Rightarrow p(Q) = 15 - 6 = 9$$
$$\Rightarrow \Pi^M \hat{=} \Pi_1 = 9^*6 - 3^*6 = 36$$

Conclusions of the Stackelberg Model

The Challenger chooses the quantity to offer in such a way that it equals the quantity that he would offer in the Cournot game. A first-mover (or early-mover) advantage exists because the Leader can set a higher output quantity and therefore has a higher profit than the Challenger. However, the Leader (initially the monopolist) does have to give up part of his profits as illustrated in the formulae (G47) and (G48) above. Conclusions of the Stackelberg game are as follows:

- The market price is reduced from 9 to 6.
- The market quantity is increased from 6 to 9.
- The sum of supplier's profit is reduced from 36 to 27 because the Challenger's entry results in a higher quantity and therefore a lower price.
- The Leader's profit is reduced from 36 to 18.

Cournot with Constant versus Increasing Marginal Cost

Having studied the classic Cournot, Bertrand, and Stackelberg game theory, I will now go a step further and calculate how Cournot would change when increasing marginal cost is assumed. While this is not a novelty, the literature rarely speaks of the impact of increasing marginal cost on prices and profits, which I illustrate herein. More important are the conclusions for corporate strategy in the coal business. The goal is to look at some intuitive market phenomena and compare them to Cournot results. All calculations have been made independent of prior research.

Most if not all natural resource markets are characterized by variable, increasing marginal cost. I have empirically shown for the coal market that this is the case. Therefore, extending Cournot to study a game with increasing marginal cost is valuable for studying any natural resource market.

My hypothesis is that the Cournot price with constant MC is lower than the price with increasing MC, all else being equal. Before calculation, I cannot hypothesize how the profit would change with increasing MC. I start below by summarizing Cournot with constant MC for comparison, followed by calculations with increasing marginal cost and attempt to confirm my hypothesis.

Cournot with Constant Marginal Cost

The results from the calculations made previously for Cournot with constant marginal cost are repeated again below.

Equilibrium quantity of each player, as per formula (G28): $q^* = \dfrac{a - c_v}{b(n+1)}$.

Market quantity in equilibrium, as per formula (G29): $Q^* = nq^* = n\dfrac{a - c_v}{b(n+1)}$.

Equilibrium market price, as per formula (G32): $p^* = \dfrac{a - c_v}{n+1} + c_v$.

Equilibrium profit of each player, as per formula (G35): $\Pi_i^*(n) = \dfrac{1}{b}\left(\dfrac{a - c_v}{n+1}\right)^2 - C_f$.

Number of players in the market at equilibrium, see (G39): $n^{\mathrm{FE}} = \left(\dfrac{1}{b}\dfrac{(a - c_v)^2}{C_f}\right)^{\frac{1}{2}} - 1$.

Cournot with Increasing Marginal Cost – Introduction

The key purpose of this analysis is to calculate the formulae for Cournot with increasing marginal cost. This exercise proves rather challenging, but with a set of basic assumptions it is possible not only to calculate the formulae for Cournot with increasing marginal cost but, as can be seen, also to determine the price premium for a market with increasing marginal cost versus a market with constant marginal cost, all else being equal.

I start again with the inverse demand function:

$$p(Q) = p(q_1, q_2, ..., q_n) = a - bQ = a - b\sum\nolimits_{i=1}^{n} q_i = a - b\left(q_i + \sum\nolimits_{j \neq i} q_j\right) \quad \text{(G49)}$$

Symmetrical firms with symmetrical cost functions are assumed. Otherwise, one could not set $q_i = q$ in the equilibrium and it would make the calculations significantly more complicated. The cost functions are therefore defined as $C_i(q_i) = C_{f_i} + c_{v_i}(q_i)$, where the marginal cost is now variable and thus depends on (or in our case increases) with the quantity q_i.

It is easy to see that the same calculations as in (G22) further above cannot be used, because the quantity q_i cannot be separated anymore in this general approach. The following lines demonstrate this problem:

$$\Pi_i = (a - bQ)q_i - C(q_i) = \left(a - b\left(q_i + \sum\nolimits_{j \neq i} q_j\right)\right) q_i - C_f + c_v(q_i) \quad \text{(G50)}$$

$$\Rightarrow \frac{\partial \Pi_i}{\partial q_i} = a - 2bq_i - b\sum\nolimits_{j \neq i} q_j - MC = a - 2bq_i - b\sum\nolimits_{j \neq i} q_j - c_v{}'(q_i) = 0 \quad \text{(G51)}$$

$$\Rightarrow a - b\sum_{j\neq i} q_j = 2bq_i + c_v{}'(q_i) \tag{G52}$$

Therefore it is not possible to separate q_i, and a general reaction function can therefore not be built. Thus, c_v cannot be kept in the calculation, because it changes to $c_v{}'(q_i)$. Since such a general approach cannot be used with knowledge and information available today, it is necessary to simplify assumptions, as discussed below.

Cournot with Increasing Marginal Cost – Simplifying Assumptions

For purposes of illustration, below I assume the cost function of each player to be

$$C_i(q_i) = \text{FC} + c_1 q_i + \frac{1}{2}c_2 q_i^2 \Rightarrow \text{MC} = c_1 + c_2 q_i \tag{G53}$$

I use this cost function because it is one of the most-used cost functions in text books. This function is one of the most commonly used because it satisfies all necessary conditions for getting a concave target function in Π_i. See also Tirole (1988) in Section 5.4 after function No. 5.8.

If potencies above 2 are attempted, for example 3, the same problem as in the general approach results, because with MC = ... + q_i^2 there is again no useful way of separating q_i in the reaction functions. Alternatively, one could change the factor by which the quantity is multiplied (in our example 1/2), but that would only slightly change the formulae and would not improve anything, as can easily be seen by using 1/3 instead of 1/2. Thus, I use 1/2 as a factor only to avoid having anything other than c_2 in front of q_i in the *MC*, to simplify.

*Calculating Each Player's Equilibrium Quantity q_i**

$$\Pi_i = (a - bQ)q_i - C(q_i) = \left(a - b\left(q_i + \sum_{j\neq i} q_j\right)\right) q_i - \text{FC} - c_1 q_i - \frac{1}{2}c_2 q_i^2 \tag{G54}$$

Optimality conditions for the correct choice of quantity

- MR − MC = 0 ⇔ MR = MC
- TR ≥ TC
- $Q = q_i + \sum_{j\neq i} q_j$ (note: only a step to be able to build the reaction function later when all q_i need to be separated)

$$\Rightarrow \frac{\partial \Pi_i}{\partial q_i} = a - 2bq_i - b\sum\nolimits_{j \neq i} q_j - \mathrm{MC} = a - 2bq_i - b\sum\nolimits_{j \neq i} q_j - c_1 - c_2 q_i = 0 \tag{G55}$$

$$\Leftrightarrow a - b\sum\nolimits_{j \neq i} q_j - c_1 = 2bq_i + c_2 q_i = q_i(2b + c_2) \tag{G56}$$

The reaction function is now built by resolving for q_i for every possible quantity of the remaining companies q_{-i}.

$$\Rightarrow R_i(q_{-i}) = q_i = \frac{a - b\sum_{j \neq i} q_j - c_1}{2b + c_2} \tag{G57}$$

The reaction functions for all companies are the same as above because all companies are symmetrical. Therefore, q_i can be set equal to q in the equilibrium and the appropriate quantity of each market player can be determined for where n equals the number of all market players.

$$\begin{aligned} \Rightarrow q = \frac{a - b(n-1)q - c_1}{2b + c_2} &\Leftrightarrow q(2b + c_2) + b(n-1)q = a - c_1 \Leftrightarrow q((2b + c_2) \\ &+ b(n-1)) = q(b + bn + c_2) = a - c_1 \\ &\Leftrightarrow q^* = \frac{a - c_1}{b(n+1) + c_2} \end{aligned} \tag{G58}$$

Calculating the Total Market Quantity Q in Equilibrium*

$$Q^* = nq^* = n\frac{a - c_1}{b(n+1) + c_2} \tag{G59}$$

*Calculating the Equilibrium Market Price p**

$$\begin{aligned} p^*(Q) &= a - bQ = a - b\frac{n(a - c_1)}{b(n+1) + c_2} = \frac{ab(n+1) + ac_2 - bn(a - c_1)}{b(n+1) + c_2} \\ &= \frac{abn + ab + ac_2 - abn + bnc_1}{b(n+1) + c_2} \\ &= \frac{a(b + c_2) + bnc_1}{b(n+1) + c_2} \end{aligned} \tag{G60}$$

Calculating the Equilibrium Profit Π for Each Player*

When calculating the equilibrium profit, the calculated formulae (G53), (G58), and (G60) above are utilized.

$$\Pi_i^*(n) = (p^* - \mathrm{MC}_i)q_i^* - \mathrm{FC}$$

$$= \left(\frac{a(b+c_2)+bnc_1}{b(n+1)+c_2} - c_1 - c_2\frac{a-c_1}{b(n+1)+c_2}\right)\left(\frac{a-c_1}{b(n+1)+c_2}\right) - \mathrm{FC}$$

$$= \left(\frac{a(b+c_2)+bnc_1 - bc_1(n+1) - c_1c_2 - c_2(a-c_1)}{b(n+1)+c_2}\right)\left(\frac{a-c_1}{b(n+1)+c_2}\right) - \mathrm{FC}$$

$$= \left(\frac{ab+ac_2+bnc_1 - bnc_1 - bc_1 - c_1c_2 - c_2a + c_1c_2}{b(n+1)+1}\right)\left(\frac{a-c_1}{b(n+1)+c_2}\right) - \mathrm{FC}$$

$$= \left(\frac{b(a-c_1)}{b(n+1)+c_2}\right)\left(\frac{a-c_1}{b(n+1)+c_2}\right) - \mathrm{FC} = \frac{(a-c_1)^2 b}{(bn+b+c_2)^2} - \mathrm{FC} \tag{G61}$$

$$\Rightarrow \Pi_i^*(n) = \frac{(a-c_1)^2 b}{(bn+b+c_2)^2} - \mathrm{FC} \tag{G62}$$

Calculating How Many Companies n Will Enter the Market

The number of companies that enter the market assuming free entry n^{FE} is calculated, starting with formula (G62) above.

$$\Pi_i^*(n) = \frac{(a-c_1)^2 b}{(bn+b+c_2)^2} - \mathrm{FC} = 0$$

$$\Leftrightarrow \frac{(a-c_1)^2 b}{(bn+b+c_2)^2} = \mathrm{FC} \tag{G63}$$

$$\Leftrightarrow \left(\frac{(a-c_1)^2 b}{\mathrm{FC}}\right)^{\frac{1}{2}} = b(n+1)+c_2$$

$$\Leftrightarrow \frac{\left(\frac{(a-c_1)^2 b}{\mathrm{FC}}\right)^{\frac{1}{2}} - c_2}{b} - 1 = n^{\mathrm{FE}} = n^{\mathrm{FE}} = \frac{1}{b}\left(\frac{(a-c_1)^2 b}{\mathrm{FC}}\right)^{\frac{1}{2}} - c_2 - 1 \tag{G64}$$

Comparison of Price with Constant versus Increasing Marginal Cost

When comparing the price with constant marginal cost versus the price with increasing marginal cost under Cournot I come to the following conclusions:

- When c_v is increased, the price in constant MC increases. In increasing MC it would also increase because $c_v = c_1$ has to be set for comparison as can be seen in the paragraphs below (Note: $c_v, c_1 \leq a$, otherwise the companies would not be in business).
- Thus, in constant MC, $p^*(Q) = \dfrac{a - c_v}{n+1} + c_v \xrightarrow{\lim c_v \to a} a.$
- Thus, on increasing MC, $p^*(Q) = \dfrac{a(b + c_2) + bnc_1}{b(n+1) + c_2} \xrightarrow{\lim c_1 \to \infty}$ undef.
- When b is increased, then the price in constant MC remains constant, and on increasing MC the price decreases.
- Thus, on increasing MC, $p^*(Q) = \dfrac{a(b + c_2) + bnc_1}{b(n+1) + c_2} \xrightarrow{\lim b \to \infty} \dfrac{a + nc_1}{n+1}.$
- When c_2 is increased, then the price in constant MC remains constant, and on increasing MC the price increases.
- Thus on increasing MC, $p^*(Q) = \dfrac{a(b + c_2) + bnc_1}{b(n+1) + c_2} \xrightarrow{\lim c_2 \to \infty} a$

Graphical Price Comparison with Constant versus Increasing Marginal Cost

After these basic comparisons, I would now like to compare the relative levels of price with constant MC versus price with increasing MC. My hypothesis was that the price with constant MC is lower than the price with increasing MC, all else being equal. In order to demonstrate that the price with constant MC is lower than the price with increasing MC, I first make use of graphs.

Figure G3 shows the simple demand curve. Figure G4 shows the cost curves for (1) the constant marginal cost case and (2) the increasing marginal cost case.

In order to compare the price with constant MC versus increasing MC, the same starting point for the two cost curves on the left side, the ordinate, is required. It is clear that $c_1 = c_v$ has to be set for this. This means, further, that there is no way that the increasing marginal cost curve lies at any point under the constant marginal cost curve, because of the steadily increasing form of MC_{var}. The logical conclusion is that MC_{var} has the higher point of intersection with a demand curve for a single firm, even if it is not possible to illustrate or specify this kind of demand curve. But

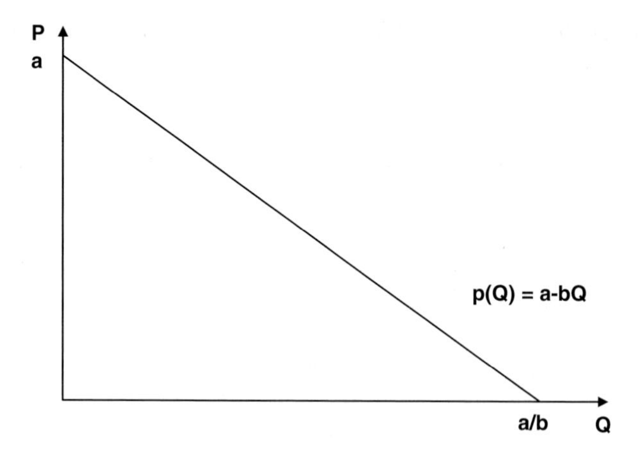

Fig. G3 Graphical figure of demand curve (Source: Author)

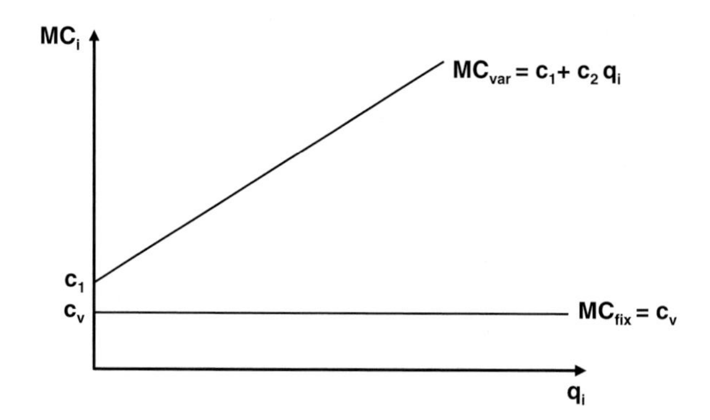

Fig. G4 Graphical figure of constant and increasing MC curves (I) (Source: Author)

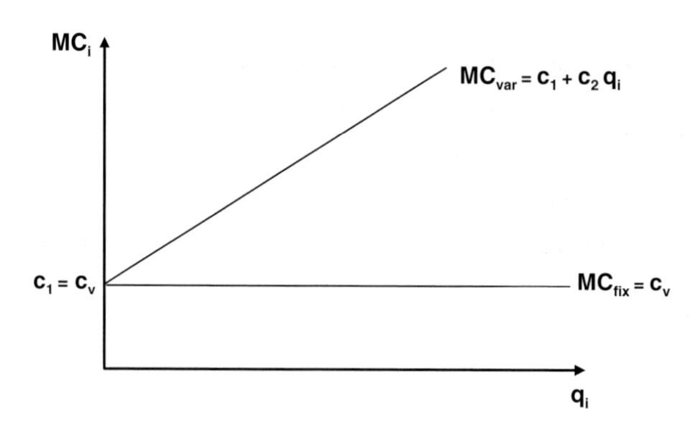

Fig. G5 Graphical figure of constant and increasing MC curves (II) (Source: Author)

this has to mean that increasing MC results in a higher price than constant MC, all else equal (Fig. G5).

Non-graphical Price Comparison with Constant versus Increasing Marginal Cost

Assuming $c_1 = c_v$, I compare $p^*(Q)_{\text{fix}} = \frac{a-c_1}{n+1} + c_1$ with $p^*(Q)_{\text{var}} = \frac{a(b+c_2)+bnc_1}{b(n+1)+c_2}$ and determine the price premium PP below.

$$\text{PP} = p^*(Q)_{\text{var}} - p^*(Q)_{\text{fix}} = \frac{a(b+c_2)+bnc_1}{b(n+1)+c_2} - \frac{a-c_1}{n+1} - c_1$$

$$= \frac{(a(b+c_2)+bnc_1)(n+1) - (a-c_1)(b(n+1)+c_2) - c_1(n+1)(b(n+1)+c_2)}{(b(n+1)+c_2)(n+1)}$$

$$= \frac{\begin{pmatrix} abn + ac_2n + bn^2c_1 + ab + ac_2 + bnc_1 - abn - ab - ac_2 + c_1bn + c_1b \\ +c_1c_2 - c_1bn^2 - 2c_1bn - c_1nc_2 - c_1b - c_1c_2 \end{pmatrix}}{(b(n+1)+c_2)(n+1)}$$

$$= \frac{ac_2n - c_1nc_2}{(b(n+1)+c_2)(n+1)} = \frac{(a-c_1)nc_2}{(b(n+1)+c_2)(n+1)} \tag{G65}$$

The question now is whether the price premium is positive or negative. To check this, the following facts should be remembered:

- $a > 0$, to have a positive starting point for the demand curve on the ordinate;
- $c_v \leq a$, otherwise the company with constant MC would not be in business at all;
- $c_1 \leq a$, otherwise the company with increasing MC would not be in business at all;
- $b > 0$, to have a negative slope of the demand curve;
- $n > 0$, to have a positive number of suppliers; and
- $c_v \geq 0$ and $c_2 \geq 0$, to have nonnegative variable costs.

It thus becomes evident that the price premium has to be positive (or zero), because the denominator is a product with two positive factors and the numerator, therefore, has to be positive to get any equilibrium in the market with a positive supply (see also $q^* = \frac{a-c_v}{b(n+1)} \geq 0$). I conclude as follows:

- The price premium for increasing MC is positive:
  $$\text{PP} = \frac{(a-c_1)nc_2}{(b(n+1)+c_2)(n+1)} \geq 0$$
- The price premium converges to 0 if n becomes large:
  $$\text{PP} = = \frac{(a-c_1)nc_2}{(b(n+1)+c_2)(n+1)} \xrightarrow{\lim n\to\infty} = 0$$

Conclusion: It has been shown, given the described assumptions, that the price with constant MC is lower than the price with increasing MC. Also, a shortage which provokes an increase in c_2 results in a higher price premium. At the same time, a reduction in the number of market participants will increase the price premium, which supports consolidation.

Comparison of Profit with Constant versus Increasing Marginal Cost

When comparing the profit with constant marginal cost versus the profit with increasing marginal cost under Cournot I come to the following conclusions:

- When c_v is increased, then the profit in constant MC and increasing MC decreases (assuming $c_1 = c_v$).
- Thus in constant MC, $\Pi_i^*(n) = \frac{1}{b}\left(\frac{a-c_v}{n+1}\right)^2 - C_f \xrightarrow{\lim c_v \to a} -C_f$.
- Thus in increasing MC, $\Pi_i^*(n) = \frac{(a-c_1)^2 b}{(bn+b+c_2)^2} - \text{FC} \xrightarrow{\lim c_1 \to a} -\text{FC}$.
- When b is increased, then both profits decrease.
- Thus in constant MC, $\Pi_i^*(n) = \frac{1}{b}\left(\frac{a-c_v}{n+1}\right)^2 - C_f \xrightarrow{\lim b \to \infty} -C_f$.
- Thus in increasing MC, $\Pi_i^*(n) = \frac{(a-c_1)^2 b}{(bn+b+c_2)^2} - \text{FC} \xrightarrow{\lim b \to \infty} -\text{FC}$.
- When c_2 is increased, or when the MC curve becomes steeper, then the profit in constant MC remains constant, and in increasing MC the profit decreases.
- Thus in increasing MC, $\Pi_i^*(n) = \frac{(a-c_1)^2 b}{(bn+b+c_2)^2} - \text{FC} \xrightarrow{\lim c_2 \to \infty} -\text{FC}$.

I now move on to the non-graphical comparison, because a complicated graphical comparison will not help much in this case. But as in discussed before $c_1 = c_v$ has to be set. In addition, the same fixed costs have to be used by setting $C_f = \text{FC}$. As a result, $\Pi_i^*(n)_{\text{fix}} = \frac{1}{b}\left(\frac{a-c_1}{n+1}\right)^2 - C_f$ can now be compared with $\Pi_i^*(n)_{\text{var}} = \frac{(a-c_1)^2 b}{(bn+b+c_2)^2} - C_f$ and the profit premium ΠP can be determined. The profit with constant MC is subtracted from the profit with increasing marginal cost.

$$\Pi P = \Pi_i^*(n)_{\text{var}} - \Pi_i^*(n)_{\text{fix}} = \frac{(a-c_1)^2 b}{(b(n+1)+c_2)^2} - \frac{1}{b}\left(\frac{a-c_1}{n+1}\right)^2$$

$$= \frac{(a-c_1)^2 b^2 (n+1)^2 - (a-c_1)^2 (b(n+1)+c_2)^2}{(b(n+1)+c_2)^2\, b(n+1)^2}$$

$$= \frac{(a-c_1)^2 \left[b^2(n+1)^2 - (b(n+1)+c_2)^2\right]}{(b(n+1)+c_2)^2\, b(n+1)^2}$$

$$= \frac{(a-c_1)^2 \left[b^2n^2 + 2b^2n + b^2 - b^2n^2 - 2b^2n - b^2 - 2bc_2n - 2bc_2 - c_2^2\right]}{(b(n+1)+c_2)^2\, b(n+1)^2}$$

$$= \frac{(a-c_1)^2 \left[-2bc_2(n+1) - c_2^2\right]}{(b(n+1)+c_2)^2\, b(n+1)^2} \tag{G66}$$

The question now is whether the profit premium is positive or negative. To check this, the following facts should be again remembered:

- $a > 0$, to have a positive starting point for the demand curve on the ordinate;
- $b > 0$, to have e negative slope of the demand curve;
- $n > 0$, to have a positive number of supplier;
- $c_v \leq a$, otherwise the company with constant MC would not be in business at all;
- $c_1 \leq a$, otherwise the company with increasing MC would not be in business at all;
- $c_v \geq 0$ and $c_2 \geq 0$, to have nonnegative variable costs.

It thus becomes evident that the profit premium has to be negative (or zero), because the denominator is a product with two squared terms and a third positive factor, and thus remains positive. The numerator is a product of one squared term that is hence positive and a second term in square brackets. That second term in square brackets is negative; thus, the numerator becomes negative (the first part is a positive product with a minus in front and the second part is a squared number with a minus in front). The Profit Premium for increasing MC is therefore negative:

$$\Pi P = \frac{(a-c_1)^2 \left[-2bc_2(n+1) - c_2^2\right]}{(b(n+1)+c_2)^2\, b(n+1)^2} \leq 0 \tag{G67}$$

That means that the profit with increasing MC is smaller than the profit with constant MC: $\Pi_i^*(n)_{\text{var}} \leq \Pi_i^*(n)_{\text{fix}}$. I have shown before that $p^*(Q)_{\text{var}} \geq p^*(Q)_{\text{fix}}$. Thus, a higher price in the case of increasing marginal cost reduces the equilibrium quantity so much that it results in a loss of profit for each player and thus for the industry.

The profit premium converges to 0 if n becomes very large. This follows logically from the fact that with an immense number of firms, the market price converges to the marginal cost.

$$\Pi P = \frac{(a-c_1)^2\left[-2bc_2(n+1)-c_2^2\right]}{(b(n+1)+c_2)^2\, b(n+1)^2} \xrightarrow{\lim n\to\infty} 0 \tag{G68}$$

The negative profit premium converges as below when c_2 becomes very large; thus when the MC curve becomes steep, as in scarce resource market would be the case.

$$\Pi P = \frac{(a-c_1)^2\left[-2bc_2(n+1)-c_2^2\right]}{(b(n+1)+c_2)^2\, b(n+1)^2} \xrightarrow{\lim c_2\to\infty} \frac{-(a-c_1)^2}{b(n+1)^2} \tag{G69}$$

In this chapter I have relaxed one major assumption of the Cournot game. I have calculated Cournot with increasing marginal cost.

***Overall implications*:** I have shown that price increases with increasing marginal cost versus constant marginal cost and that profit decreases. It has been shown that in scarce resource markets where the MC curve becomes steeper prices increase, thus confirming the hypothesis and market intuition. More interestingly, it has been shown that firms should strive for a flat MC curve as their profits decrease, the steeper their MC curve becomes. Symmetrical firms had to be assumed, but in real life this becomes especially relevant for the marginal cost producers in a scarce market whose profits decrease the steeper their MC curve is. However, firms that produce with costs below the cost of the marginal cost producer are very interested in scarce markets because the resulting price increase goes straight to their bottom line. As before, the market tends toward consolidation as fewer competitors also increase profits in markets with increasing marginal cost.

Index